New City Landmark—MIXED-use Architecture

城市新地标——综合体建筑

深圳市艺力文化发展有限公司 编

华南理工大学出版社
SOUTH CHINA UNIVERSITY OF TECHNOLOGY PRESS
·广州·

图书在版编目（CIP）数据

城市新地标： 综合体建筑 = New city landmark: mixed-use architecture: 英汉对照 / 深圳市艺力文化发展有限公司编 . — 广州：华南理工大学出版社, 2015.2
ISBN 978-7-5623-4401-8

Ⅰ．①城… Ⅱ．①深… Ⅲ．①综合建筑－建筑设计－作品集－世界 Ⅳ．① TU24

中国版本图书馆 CIP 数据核字（2014）第 211258 号

城市新地标 —— 综合体建筑
New City Landmark — Mixed-use Architecture
深圳市艺力文化发展有限公司 编
_____

出 版 人：	韩中伟
出版发行：	华南理工大学出版社
	（广州五山华南理工大学 17 号楼，邮编 510640）
	http://www.scutpress.com.cn E-mail: scutc13@scut.edu.cn
	营销部电话：020-87113487 87111048（传真）
策划编辑：	赖淑华
责任编辑：	陈　昊　黄丽谊
印 刷 者：	深圳市汇亿丰印刷科技有限公司
开　　本：	595mm×1020mm 1/16 **印张**：25.5
成品尺寸：	248mm × 290mm
版　　次：	2015 年 2 月第 1 版 2015 年 2 月第 1 次印刷
定　　价：	398.00 元

_____

版权所有　盗版必究　　印装差错　负责调换

扫描二维码，开启电子阅读体验，海量优秀作品随心看！

国际视野新站点 | 案例丰富新颖 | 访谈顶尖设计 | 挖掘新锐设计师 | 国际设计界缩影

ACS 创意·空间 — ABOUT

十年专注于建筑、室内、景观和平面设计，业务横跨图书出版、发行、文化传媒、品牌运营及艺术品市场等多个经营领域，Artpower 自版发行 600 多本图书，收揽全球顶尖设计公司和设计师近 40 000 套优秀原创作品（不断更新ING）。

ACS 整合 Artpower 线上资源，推荐前沿创意理念、概念性设计思维；发布创意赛事活动；组织设计大师访谈；展示新锐设计师作品，推介设计项目；提供私人定制出版和众筹出版等服务。与国际设计团队对接，在全球范围内打造专业设计师展示和交流平台。

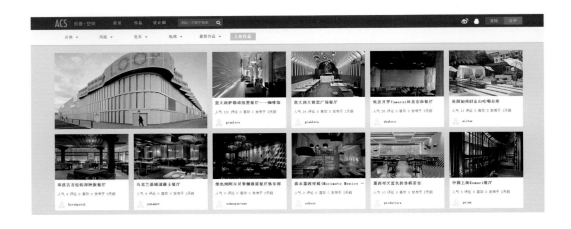

我们能做什么？ — HOW

注册成为网站会员，做主个人网页，独享会员特权；上传个人作品，展示设计理念，交流成长，互通合作契机。

登录浏览，尽享 40000+ 海内外设计大师作品；建筑、室内、景观、平面、产品、环境设计等分门别类，轻松导航，应有尽有。

挖掘新锐设计师 — DESIGNERS

· ACS 线上展厅
我的 ACS 我做主！设计师可以尽情发表自己的作品，让世界各地的设计师共同关注你的成长！

· 设计师发布会
如果你还在为身为"新人"的标签所困扰，ACS 展示平台只有"新锐设计师"。把曾经因各种原因被否掉的方案重新发表出来，也许你就是那个我们要找的设计师！

· ACS 把设计师的项目推送给全世界，设计无国界，一起关注和交流！

私人定制出版 — PRIVATE

ACS 创意空间联营平台为您提供私人定制出版服务。

线下俱乐部 — ACTIVE

ACS 创意空间俱乐部，不定期邀请国内外顶尖设计师，举办各种创意设计讲座、创意沙龙等，分享天马行空的有趣创意，是思维碰撞、灵感横溢的场所，是趣味相投、惺惺相惜的交友平台，也是企业品牌的展示空间。有机会成为线下俱乐部盟主！

DESIGN FOR DESIGN

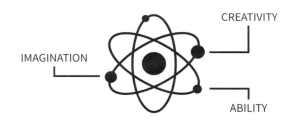

IMAGINATION · CREATIVITY · ABILITY

设计 · 杂志 · 中英文 · 双月刊
Artpower Creative Space（ACS）创意空间（245mm×325mm · 168 页 · 68CNY）

《ACS 创意空间》杂志是 Artpower 倾力打造的高端空间设计专业期刊。中英双语，全球同步发行；单期发行量逾万册，更有黎巴嫩等国家的专售版；装饰行业至佳交流平台，传播设计新锐资讯；高端空间设计专业期刊，发布国际最优秀室内设计师和建筑设计师的最新作品。

深圳市艺力文化发展有限公司
艺力国际出版有限公司（香港）
深圳市艺力文化发展有限公司北京分公司
深圳市艺力文化发展有限公司厦门分公司

出版合作 / 广告合作：rainly@artpower.com.cn（王小姐）
作品投稿：artpower@artpower.com.cn（莫小姐）

艺力 ACS 创意空间
扫描即可关注！

PREFACE /序言

In the recent two years, urban complexes, with commercial, office, residential, dining, entertainment and transportation spaces as one, has been the main trend in urban development as the multifunctional complexes with high efficiency.

Due to its strong market demand and fierce competition, the developers have been aware of the importance of design. They look forward to high-quality design urgently since an advanced design will lead the whole project more competitive in all aspects at the market. In addition, to complete a high-quality mixed-use project, it needs the developers and architectural designer's effective cooperation. The developers should have high efficiency and speed of real estate and commercial operations while the designers should have effective design and quick response to party A's requirements.

In the modern society, with the prevalence of online shopping, in-store purchase has been strongly affected. But the design of amphibianArc focuses on enhance the consumers' shopping experience completely, to form the advantages in the consumers' cognition. From the architectural design point of view, there are several ways to achieve this:

•To create comfortable shopping environment. Since Hanhai Huanghe Road Mixed Use Development is located in an ancient, sombre and gloomy Northern city, the designers have used multi colors and symbolic decorative patterns, which have made the project bright and comfortable, offering consumers a pleasant and comfortable shopping environment.

•To create thematic shopping environment. The thematic shopping environment can add interest, esthetics and additional room for imagination throughout the shopping process. Hanhai Beijin Mixed Use Development has used the cultural symbol representing "Long Pan Hu Ju", with the continuous streamlined design, making the flowing line on the building curtain wall echo the sidewalk, stream and transit line around the site. The design of Hangzhou Wonder Mall and Bengbu Paramount International Cultural Commercial and Trade Square reflect the specific environmental requirements of the wedding celebration theme, providing the proprietors with opportunities to get involved in the wedding celebration industry chains.

•To creat interactive and participatory shopping environment. Hangzhou Wonder Mall allows consumers to enjoy boating and appreciate waterfront commercial scenery on the two sides of the canal. There is even an electronic screen which can interact with the smart phones.

All in all, the design focusing on the theme and shopping experiencing will become a trend in the future commercial real estate planning.

　　近两年，集商业、办公、居住、餐饮、文娱和交通等为一体的城市综合体作为一种多功能、高效率的组合形式，成为都市发展的主要趋势。

　　由于市场需求旺盛且竞争激烈，开发商对设计重要性的意识越来越强烈。他们迫切期待着优质的设计，因为一个超前的设计会促使整个项目在各方面都更具市场竞争力。另外，要成就一个优质综合体项目，需要开发商和建筑设计师共同的高效率合作。开发商应该有高效的地产开发和商业运作速度，设计师也要具备高效的设计速度和对甲方要求的回应速度。

　　现代社会网购的流行，使得实体店消费受到不小的冲击。然而amphibianArc的设计注重全面提升消费者的购物经验，在消费者认知当中形成实体店消费的优势。从建筑设计的角度而言，有几种方式达到这样的优势效果：

　　•营造舒适的购物环境。在河南瀚海黄河路商业综合体的设计中，由于项目所处的古老北方城市略显灰暗而阴沉，设计师则运用了颜色及象征式的装饰花纹，使项目变得明亮舒服起来，给消费者一个舒适惬意的购物环境。

　　•构造主题性购物环境。主题性的购物环境能够在人们整个购物过程中添加趣味性、观赏性和附加的想象空间。瀚海北金商业综合体项目就借用了象征"龙盘虎踞"的文化符号，利用连续的流线型设计，建筑幕墙上的流线也同基地附近的步道、水流和交通动线相映成趣；杭州万象天成·运河汇城市综合体和蚌埠百乐门国际文化经贸广场的设计均迎合了婚庆主题的具体环境要求，为业主提供参与当前炙手可热的婚庆产业链的机会。

　　•创造具有互动性和参与性的购物环境。杭州万象天成·运河汇城市综合体可以让消费者荡舟于运河欣赏两岸的滨水商业风光，更设有可以与智能手机互动的电子显示屏幕。

　　总而言之，强调主题性和购物经验的设计将成为未来商业地产规划不可或缺的一种趋势。

Contents / 设计师名录

板桥圆顶中心
Pangyo Dome 002

弹子石中央商务区
Danzishi Central Business District 008

迈阿密海滩广场
Miami Beach Square 016

杭州万象天成·运河汇城市综合体
Hangzhou Wonder Mall 026

三颗珍珠科技广场
The Three Pearls Scitech Plaza 032

双景坊
DUO 038

红星美凯龙北京旗舰店
Hongxing Macalline Furniture Beijing Flagship Store 046

西安利君时尚购物中心
Xi'an Lijun Mall 056

杭州文娱体育中心
Hangzhou Civic Sports Center 062

Matn Point
Matn Point 068

南京14工作室
Nanjing Studio 14 072

SRE 河滨公寓
SRE Riverside Apartment Complex 080

长沙世茂广场
Changsha Shimao Plaza 086

深圳森林广场
Shenzhen Jungle Plaza 092

欧罗巴市
Europa City 098

上海五洲国际广场
Shanghai Wuzhou International Plaza 108

杭州新天地商务中心
Hangzhou Xintiandi Commercial Center 114

苏州大运城 & 苏州花样城
Suzhou Dayun City & Suzhou Huayang City 122

文化综合体
Cultural Complex 136

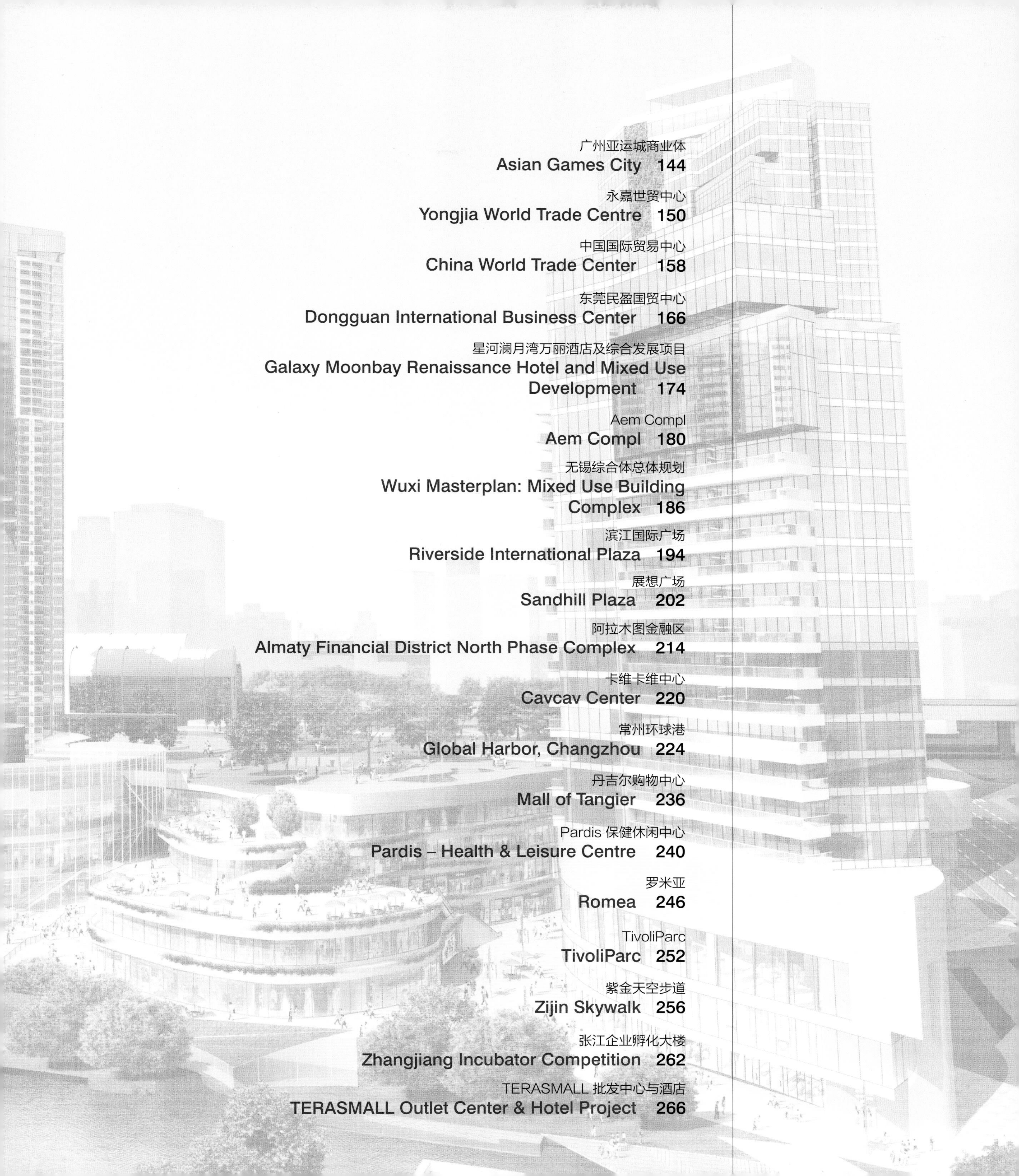

广州亚运城商业体
Asian Games City 144

永嘉世贸中心
Yongjia World Trade Centre 150

中国国际贸易中心
China World Trade Center 158

东莞民盈国贸中心
Dongguan International Business Center 166

星河澜月湾万丽酒店及综合发展项目
Galaxy Moonbay Renaissance Hotel and Mixed Use Development 174

Aem Compl
Aem Compl 180

无锡综合体总体规划
Wuxi Masterplan: Mixed Use Building Complex 186

滨江国际广场
Riverside International Plaza 194

展想广场
Sandhill Plaza 202

阿拉木图金融区
Almaty Financial District North Phase Complex 214

卡维卡维中心
Cavcav Center 220

常州环球港
Global Harbor, Changzhou 224

丹吉尔购物中心
Mall of Tangier 236

Pardis 保健休闲中心
Pardis – Health & Leisure Centre 240

罗米亚
Romea 246

TivoliParc
TivoliParc 252

紫金天空步道
Zijin Skywalk 256

张江企业孵化大楼
Zhangjiang Incubator Competition 262

TERASMALL 批发中心与酒店
TERASMALL Outlet Center & Hotel Project 266

北京西南饭店
South West Hotel 272

青岛总规划
Qingdao Master Plan 276

上海文化金融城
Shanghai Cultural and Financial City 280

方正
THE:SQUARE ³ 286

法拉盛综合体
Flushing Commons 298

西南滨水码头
The Wharf Southwest Waterfront 302

天津解放南路商业中心
Tianjin Jie Fang South Road Commercial Center 314

九方购物中心
9 Square Shopping Center 324

宁波来福士广场
Raffles City Ningbo 338

昆士伏广场
Kuntsevo Plaza 346

番禺万博 CBD 商业广场
Panyu Wanbo CBD Commercial Plaza 352

东莞长安万达广场
Dongguan Chang'an Wanda Plaza 360

杨树浦发电厂综合改造项目
Yangshupu Power Plant Renovation Project 366

宁波隆兴广场
Ningbo Long Xing Plaza 372

银川阅海万家规划
Yinchuan YueHaiWanJia Development 378

银川正源街规划
Yinchuan Zhengyuanjie Development 382

成都阳光新业中心
Yangguang Chengdu 386

设计师名录
Contributors 392

• New City Landmark — Mixed-use Architecture

Seoul, Korea
Pangyo Dome
板桥圆顶中心

Architect · Laguarda.Low Architects, LLC
Client · RS Macalline
Area · 678,450 m²

The dramatic structure of the Pangyo Dome proposes an unashamedly new, modern heart for the city. The Dome is a physically interlocking device, a series of public uses woven into a web that formally integrates the surrounding building units into a structural whole.

This project for the commercial heart of the new city of Pangyo has two goals: to provide a world class entertainment and retail district for the booming urban area, and to create a memorable visual identity for the city. This commercial center includes seven levels of retail and entertainment podium, as well as a series of semi-independent office buildings of varying heights, and a hotel and service apartments.

The ground plane of the Dome is designed as a continuous plaza garden, laid out in the form of a distinct cruciform that links all areas of the project. Here, the green corridors proposed by the city are brought through the project together with a series of

New City Landmark — Mixed-use Architecture

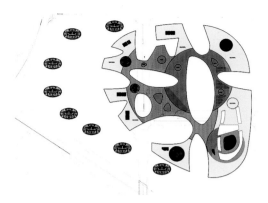
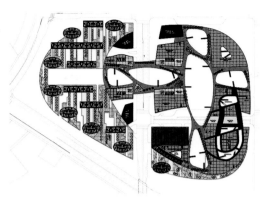
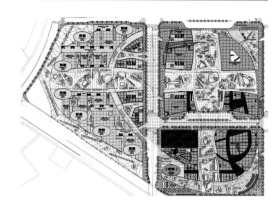

water forms, bridges, gardens, and plazas. A series of vast parks link together the entire development, containing relaxing and refreshing green areas for entertainment and amenities.

The formal strength implied by the scale and singular visual power of the unifying dome is complemented by the soft forms of the buildings, expressed both in shape and materiality. The exoskeletal roof is punctuated by voids, carved and formed as if by a river, creating eroded shapes. The idea of these voids symbolically smoothed over time by the bustling crowd in the new city center.

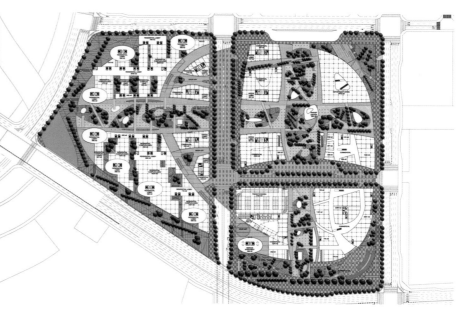

　　板桥圆顶商业中心的动态结构使之成为城市当之无愧的现代化新核心。整个圆顶商业中心采用联锁交错式设计，一系列公共场所交织成网，与周边建筑形成一个整体结构。

　　该项目位于韩国首尔南部的板桥（Pangyo）新城的核心商业区内，旨在实现两个设计目标：为这个飞速发展的城区提供一个世界级的娱乐与购物区，并为整个城市建立一个令人难忘的视觉标识。商业中心包括七层的商业和娱乐区，一系列高低各异的半独立式写字楼，另外还设有一个酒店和服务中心。

圆顶商业中心的地面层设计为一个连续的广场花园，呈独特的十字形，将项目各部分进行衔接。在这里，城市的绿色走廊连同一系列水景、小桥、花园和广场等，都与本项目相连。一系列广阔的公园将整个开发项目联系起来，为娱乐设施提供休闲绿化区域。

这个浑然一体的圆顶商业中心以其规模和统一的视觉形象产生了一种力量感，作为补充，建筑则从外形和选材上采用柔和的设计手法。外骨架屋顶以间断的空间点缀，仿佛被一条河穿过而冲蚀，象征在这个新的城市中心以熙熙攘攘的人流掩饰时光的流转。

New City Landmark — Mixed-use Architecture

New City Landmark — Mixed-use Architecture

Architect · 10 DESIGN
Client · Chongqing WOKI Real Estate,
Chongqing Energy Investment Real Estate,
Sinosteel Investment Group,
Zhongxun Group
Area · 800,000 m²
Program · Office, Five Star Hotel, Serviced Apartments, Retail

Chongqing, China

Danzishi Central Business District

弹子石中央商务区

Four corporate investors have come together to develop the 800,000 m² Danzishi Central Business District with a common vision to create a premium business and retail destination for this dynamic new district in the heart of Chongqing in China.

The project sits centrally within the new Danzishi Central Head Quarters in Chong Qing and is located on the banks of the Yangtze River. This mixed use development is one of three focal projects in the city and aims to create a high-end business and retail district. The project covers 800,000 m²; the ranges of functions include cultural, sport, hospitality and entertainment.

The design brief was to create a cohesive solution that was both architecturally innovative and sympathetic to local context and planning constraints. It was important to reflect the rich topography and culture of Chongqing.

The steep site, which slopes down to the Yangtze River, is reminiscent of the city's traditional stepped streets and buildings that faced onto the river.

A 70m level difference across the site enabled the creation of multiple levels, each activated by street frontage and each being mutually beneficial in activating upper and lower levels.

By taking advantage of this topographical feature, the multi-level circulation diagram feeds into riverside foot traffic along the promenade and a light rail station located at the top of the site, from which pedestrians are drawn into by the elevated link bridges.

由四家投资企业联合开发、总面积达 800 000 m² 的重庆弹子石中央商务区，以打造国际级商业与零售为共同愿景，在重庆中心地带创建这片广阔新城。

项目位于重庆市长江沿岸弹子石新中央总部中心地带，为该市力争区域经济强化发展的三大热点项目之一，目的是要打造成为一个高端商业与零售的集中地，提供共 800,000 ㎡ 的文化、体育、酒店及娱乐空间。

设计任务是要根据实际环境及规划限制，以创新及人性化的建筑设计提供一个新颖而具

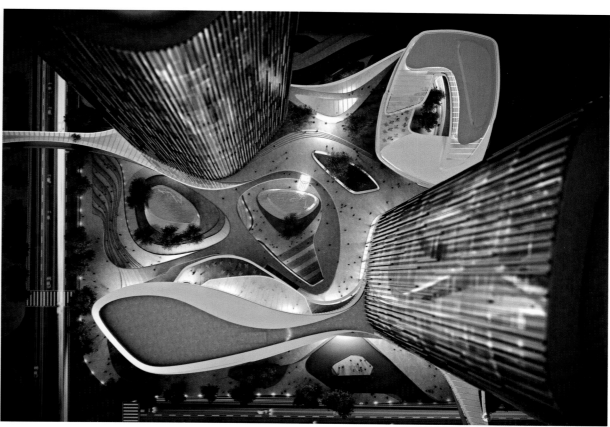

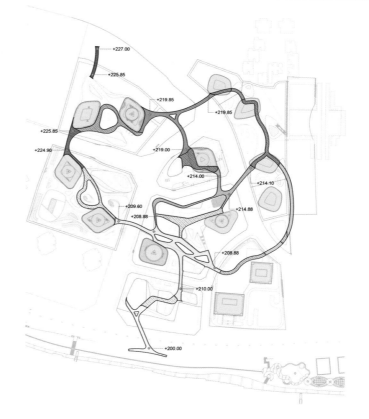

凝聚力，又能突显重庆丰富地形及文化的理想方案。

基地地势陡峭，呈向下延伸的坡道，直通长江岸，这不仅体现出重庆典型的地形特征，也令人联想起传统的楼梯街和面朝长江的旧时建筑。

70m地势高差贯穿整个开发区，打造了多重零售层，由临街面激发活力的商业业态创造了得天独厚的机遇，让每一零售层都可通过与上、下层面的互动而实现双赢。

设计方案利用这种主要的地形特征，使具有多重标高的商业裙楼价值倍增。此外，在低处江边漫步的客流将进一步活化这种多层次的零售业态，而且，位于基地最高点的轻轨站也经由高架链桥将人流源源不断地引进项目开发区的核心地段。

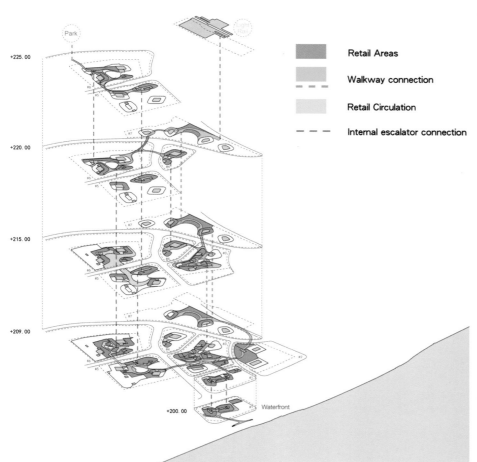

• New City Landmark — Mixed-use Architecture

Architect · BIG (master plan and architecture), West 8 (public realm), John Portman & Associates (hotel), Fentress (convention center), Revuelta Architects (local)
Client · Portman Holdings, CMC, Bal Harbor Shops, Cirque du Soleil
Program · Convention, Hotel, Cultural, Exhibition

Miami, USA

Miami Beach Square

迈阿密海滩广场

BIG unveils plans to create a new civic heart for Miami Beach by redeveloping the Convention Center. BIG together with West 8, Fentress, JPA and developers Portman CMC proposes Miami Beach Square as the centerpiece of their 21 hm² Convention Center.

Their mission is to bring Miami Beach back to the Convention Center and to imagine an architecture and an urban space unique to the climate and culture of Miami Beach.

Nearly 50% of the site is open space at grade — open lawns, shady paths and plazas, parks and gardens that forms an archipelago of urban oases throughout the site. At the heart of it the designers introduce a central square to become the pivoting point of the entire neighborhood. Becoming the front door to the convention center, and the convention hotel, a front lawn to the revitalized Jackie Gleason Theatre, a town square for the city hall, an outdoor arena for the Latin American Cultural Museum, and the red carpet for the big botanical ball room: the Miami Beach Square.

The square creates a series of intuitive connections across the site — a diagonal that connects the Soundscape to the Botanical Gardens and Holocaust Memorial. A north-south connection joins the Collins Canal to Lincoln Road and naturally channels the flow of convention visitors to the liveliness of Lincoln Road. A green network of public spaces that stitches together all of the adjacent neighborhoods — formerly separated by the convention center — in to a complete and coherent community — for both visitors and residents. All public programs — old and new come together on the square. All great cities have a great square – this will be Miami Beach Square.

While the Square is the center of the district, it is just a piece of an additional extensive network of diverse green spaces, totaling nearly 27 shady, friendly and usable acres and linked to the other great public spaces in Miami Beach .

By popular demand they have found a way to preserve and enhance the architecture and programming of the Jackie Gleason Theatre. By making it all public at the street level – opening up lobbies, restaurants and cafes on all sides – they make the Gleason a lively centerpiece in this new neighborhood. Towards the Square they propose to extend the fly tower with a performing arts center with various spaces for rehearsal and offering a visual connection to the public. Adjacent to the Jackie Gleason Theater sits the new Latin American Cultural Museum consisting of a base of public programs opening up on the square. The building form creates a covered shaded event space on the square blurring the transition between inside and outside.

Rather than being the hermetic mono programmatic box that the Miami Beach Convention Center is today – a single program at the size of an urban block – we propose to consider the Convention Center an actual urban block complete with different programs – grown together to form a continuous architecture. A gradual transition from public to private and from cultural to civic – conference to residential turns a stroll around the block into an experience of continuous variation. Along the entire west adjacent to the various gardens and the new square — the main entrances to the convention Center and Conference Center occupies the ground. The hotel lobby spans the entire south elevation in continuation of the Convention Center lobby. The Hotel facade as pulled back forming a cascade of terraces for the south facing hotel rooms – decreasing the perceived height seen from the Gleason.

New City Landmark — Mixed-use Architecture

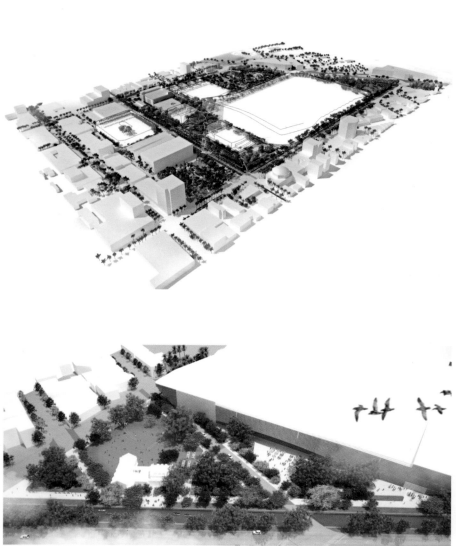

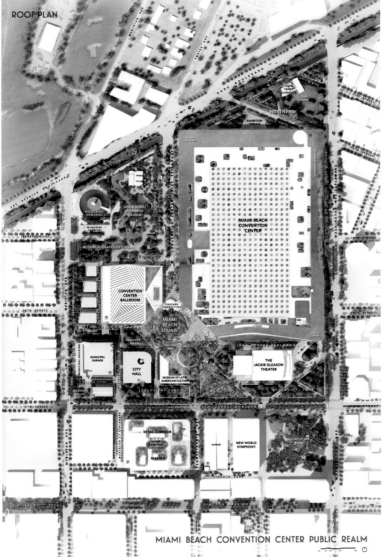

ROOF PLAN

MIAMI BEACH CONVENTION CENTER PUBLIC REALM

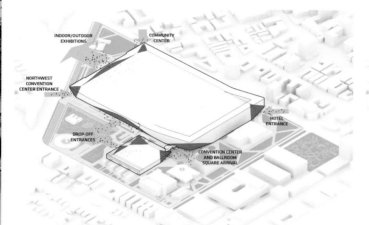

CORNER CUTS

The corners are cut off the building masses, creating shaded patios and seamless connections between indoor and outdoor spaces.

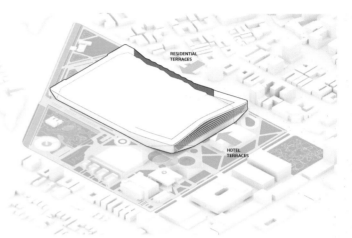

PUSH BACK

The building mass is pushed back to reveal terraces that define the residential and hotel programs.

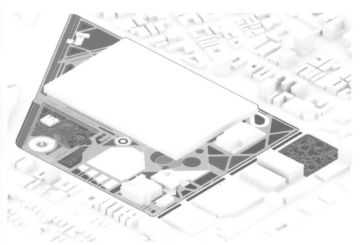

NEW LANDSCAPE

A green urban landscape provides an inviting context to the Convention Center, connecting it to the lush Miami Beach Botanical Gardens and the Soundscape Park to the south.

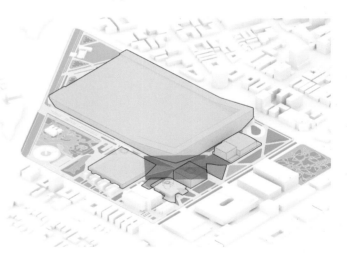

SQUARE ACTIVATION

The New Miami Beach Convention Center, the Jackie Gleason Theater, the Museum of Latin American Culture, City Hall, and the Convention Center Ballroom all sit on the square. They activate the square with civic, cultural and recreational life 24 hours a day.

BIG 设计公司运用重整会议中心的手法制定了新的迈阿密海滩民用中心计划。BIG 与 West 8、Fentress、JPA 及开发者 Portman CMC 共同提议迈阿密海滩广场应作为这 210 000 m² 的会议中心的核心部分。

我们的任务是将迈阿密海滩变回会议中心，变回设想中的建筑风格，即一个享有迈阿密海滩独一无二的气候和文化的都市空间。

将近 50% 的场所在同一水平面上，为开放式，敞开的草坪、曲幽的小径、广场、公园、花园，这些构成了整个场所中的一个都市绿洲群岛。建筑的中心，我们引进了一个中央广场使之成为整个地区的轴心。前面的草坪作为进入会议中心和会议酒店的前门，重拾活力的杰基·格里森影院拥有一块前庭草坪，市政厅拥有一个商业广场，拉丁美洲文化博物馆拥有一个室外竞技场，大型植物球形大厅拥有一方红色地毯：迈阿密海滩广场。

该广场创建了一系列横穿这片土地的直观连接，一条斜线将音景和植物花园、大屠杀纪念碑连在一起。一个南北的连接使科尔内运河与林肯路相连，河渠很自然的在会议厅来访者面前流动，给林肯路带来勃勃生机。一个绿色的公共空间网络将临近的场地拼接起来——原先被会议中心割裂——成为了来访者和居住者完整和谐的社区。所有公共项目，不管旧的和新的，共同聚集在这个广场。每一个大都市都会有一个大广场，迈阿密海滩也不例外。

由于广场是该地区的中心，广场仅仅是各种各样绿色空间网络的一个延展，大概总共有109 269m² 阴凉、良好的可用面积，与迈阿密海滩其他大型公共场所相连通。

按照一般要求，设计师发现了一种方法可以保留并改善杰克·格里森影院的建筑风格和建筑类型。通过将之在街道水平上完全开放所有边缘上的大厅、饭店、咖啡厅，使格里森成为这片新土地上活力四射的场所。面向广场设计师提议运用一个表演艺术中心延长舞台塔，这个

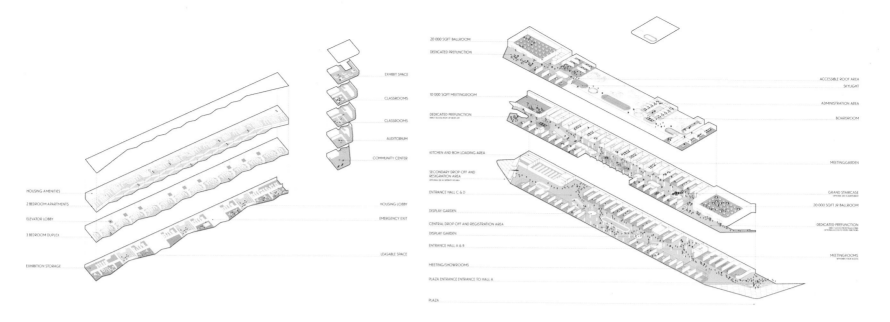

中心拥有不同的场所可供排练，也可以作为一个连通公共场所的视觉连接。临近杰克·格里森影院，坐落着一座新的拉丁美洲文化博物馆，由一个开放性广场的基本公共项目组成。建筑的外观为广场上的活动场所提供了一席遮挡，模糊了内外的转换。

与现在迈阿密海滩会议中心的密封单一项目建筑盒子不同，一个城市大楼这样规模的单一项目，我们提议将会议中心看作一个带有不同项目类型的真正的城市大楼，共同发展为一个连续性的建筑风格。渐渐地由公共场所转变为私人场所，从文化场所转变为民用场所，从会议场所转变为居住场所。沿着整个临近各式花园和这个新广场的西方会议中心的主要入口，会议中心占用了这片土地。酒店的大厅横跨了整个会议中心大厅结构中的南高地。酒店的正面向内倾斜，变成了朝南酒店房间平台的巨大帘幕，减少了从格里森角度的看到高度。

• New City Landmark — Mixed-use Architecture

Hangzhou, China

Hangzhou Wonder Mall

杭州万象天成 · 运河汇城市综合体

Architect · amphibianArc
Area · 631,806 m²

The project will be developed among decommissioned thermal power station. Conveniently located in Gongshu District, Hangzhou, China, the project is only 6.5 km from downtown and 9.5 km from the world famous West Lake.

The rapid urban development has dramatically changed the city, instead of demolishing the aging architecture, the smokestack and small scale factories are preserved. By respecting the autonomy and authenticity of the historic site, the development integrates the culturally conscious preservation with a commitment to high level solutions to embody creative tension, cultural continuity and sustainability.

New City Landmark — Mixed-use Architecture

Gongshu District is the starting point of the Beijing-Hangzhou Grand Canal, the project is designed based on the theme of canal. The figure 8-shaped waterway throughout the project connects the entire complex, with local features of bridges and boats in the historic Jiangnan area, will reproduce the prosperity of the canal city views of the past.

Visitors can enjoy a cruise along the canal to admire the beauty and uniqueness of Wonder Mall. On top of the canal, the skyscreen creates a constantly varying appearance, reflecting the inner dynamism of the shopping complex.

本案建在杭州热电厂旧址上，位于中国杭州拱墅区，离市区仅 6.5km，距离世界闻名的西湖也只有 9.5km。

快速的城市化发展令城市发生了巨大的变化，古老的建筑例如烟囱和小规模的工厂并未被摧毁，而是被保留了下来。为了尊重这一具有历史意义的场址，设计整合了文化保护意识与高水平的解决方案，体现了创造的张力和文化的连续性及可持续性。

拱墅区是京杭大运河的起点，本案以运河作为设计主题。8 字形的河道贯穿整个项目，将商业部分连接起来，配以江南水乡独有的桥和船，再现了运河两岸过去的繁荣景象。

游客可以乘船在运河游览，感受万象天成综合体的独特魅力。运河顶部的天幕，反射出购物综合体内部的动态，创造出不断变化的景致。

• New City Landmark — Mixed-use Architecture

New City Landmark — Mixed-use Architecture

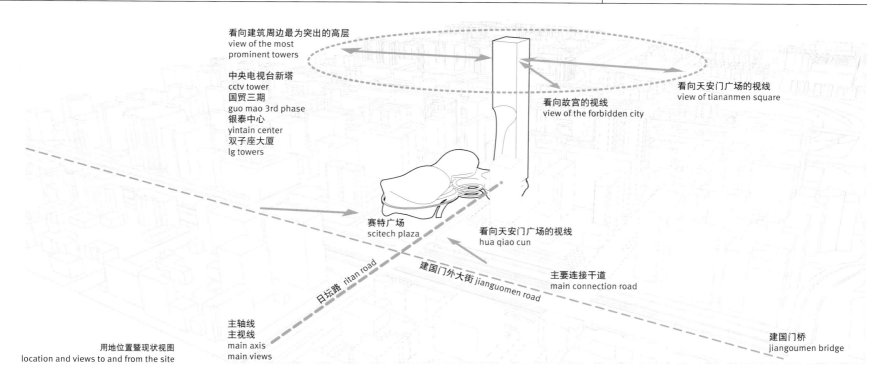

Architect · COOP HIMMELB(L)AU, Wolf D. Prix & Partner ZT GmbH
Program · Retail, Hotel, Residential, Entertainment
Renderings · COOP HIMMELB(L)AU

Beijing, China
The Three Pearls Scitech Plaza

三颗珍珠科技广场

"The development of architecture is furthered by strategies comprising of and searching for lines and fields of possibilities tied together by chance, anti-logic and unconventional thinking.

The coincidence of systems – both as built space and as media space – becomes the basis for new designs and projects, the rubber grid as the idea of a dynamic network of designs for cities like clouds." Wolf D. Prix said.

Three special stone-shaped buildings, like Pearls symbolizing wealth, luxury, uniqueness and exclusivity, form the basis of the design and house the new shopping functions as well as serve as a base for the new Scitech Tower. The Pearls are connected with each other through strands of circulation to form a continuous shopping landscape.

The new Scitech Tower is placed atop one of the pearls as a landmark for the complex, housing the Scitech offices as well as the five star hotel with two level sky bar and spa. In addition to the department store the complex also houses a 3D cinema and arcade on top of one of the Pearls. An exterior ramp with exclusive café offers an unparalleled perspective in the urban context as spectator from above.

Outside around the three Pearls forms the landscape is formed by a pattern derived from energy fields which include water, light, green and trees. The components of the landscape serve also as wayfinding elements through the Scitech complex, directing the flow of pedestrians to the building entrances.

In order to allow for a phased construction on the densely constructed site the Three Pearls are located in accordance with the former locations of the previously existing Scitech Plaza, Scitech Place and Scitech Hotel, allowing a seamless and easy three-step phasing of construction while never looking incomplete. This layout of distinct building volumes also keeps the memory of the old department store, while reformulating its new expression for the 21 century, with the exclusive nature of the luxury of the goods inside the mall communicated by its architectural expression.

设计战略能进一步加强建筑体的开发，包括机会、反逻辑和非常规思维一起形成的各种可能性。

"由于该建筑既作为建筑空间又作为媒体空间，体系的一致性成为新设计和工程的基础，橡胶网格是动态网格设计的意象，像云般。"Wolf D. Prix。

三个特殊石头形状的建筑物像珍珠，象征着财富、奢侈、唯一性，它们是该设计的基础，提供购物功能，作为该新科技大楼的底部。珍珠通过一条条循环路线相互联系，形成一片连续的商铺景致。

其中一颗珍珠作为该综合体的地标，新科技大楼位于其顶部，大楼包括科技办公室以及

拥有两层天空酒吧和水疗中心的五星级酒店。除了百货商店，综合体还有一个3D电影院和位于珍珠顶部的商城。一个外部坡道上有单独一家咖啡厅，在那有着无可比拟的观景视角点，游客能尽情俯瞰城市夜景。

在三颗珍珠之外的景观形式是来自能量场的一个模式，包括水、光、植被。景观的组成部分同样作为科技综合体的指路元素，指引行人到建筑入口。

为了在这个密集建设场址上能进行分期施工，三颗珍珠与之前存在的科技大厦、科技广场和科技酒店保持一致，这允许连续而简单的三期逐步建设，虽然施工期间看起来不完整。这个与众不同的建筑体的布局也保持了老百货商店的面貌，而商场内奢侈品的珍贵性通过建筑体表达出来。

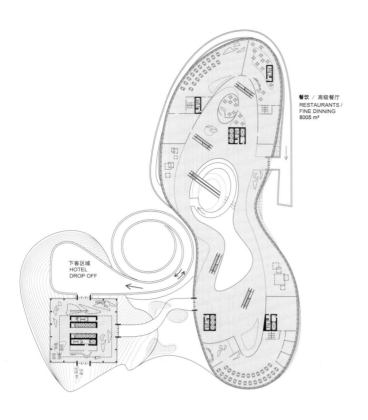

餐饮 / 高级餐厅
RESTAURANTS / FINE DINNING
8005 m²

下客区域
HOTEL DROP OFF

6 层 / +22.00m
level 6 / +22.00m

平面图 floor plan | 1:1000

总平面图 siteplan | 1:1000

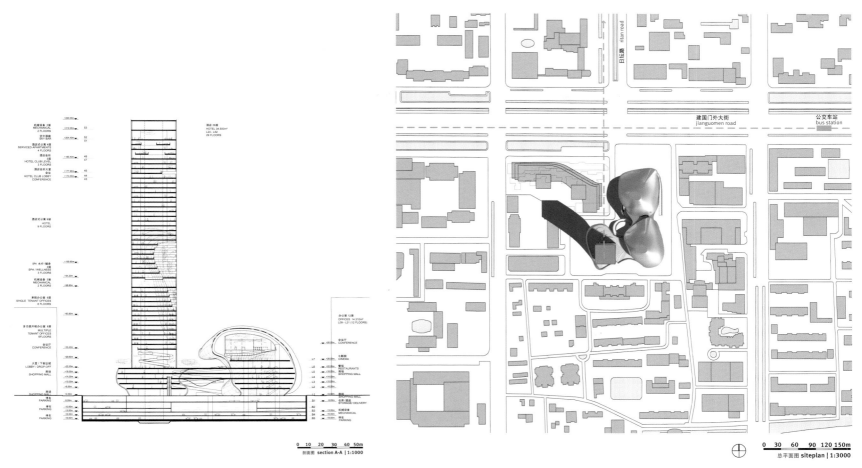

- New City Landmark — Mixed-use Architecture

New City Landmark — Mixed-use Architecture

Architect · Buro Ole Scheeren

Singapore

DUO

双景坊

The design for this Malaysian-Singaporean joint venture actively engages the space of the surrounding city to form a new civic nucleus in Singapore's modern metropolis. The two towers are not conceived as autonomous objects, but defined by the spaces they create around them.

Singapore consistently ranks as one of Asia's most livable cities. However, it is increasingly dominated by isolated individual towers that favor exclusion over social connectivity. The prescribed zoning confronts the project with a dual dilemma: it splits the site into two separate pieces, and leaves large, bulky footprints for the resulting tower envelopes which risk overpowering the surroundings and the intimate scale of the adjacent historic Kampong Glam district.

The design for DUO subtracts circular carvings from the allowable building volumes in a series of concave movements that generate urban spaces — a kind of "urban poché" that coopts adjacent buildings and symbiotically inscribes the two towers into their context.

By generating the massing through a subtractive process, the elevations of the new towers are reduced to slender profiles. Vertical facades rise skywards along the adjoining roads, while a net-like hexagonal pattern of sunshades reinforces the dynamic concave shapes. The DUO of tower volumes is further sculpted to feature a series of cantilevers and setbacks that evoke choreographed kinetic movements of the building silhouettes.

The buildings dematerialize as they reach the ground to provide a porous permeable landscape traversing the site. Leisure zones and gardens act as a connector between multiple transport hubs and establish a flow of tropical greenery and lively commercial activity, accessible to the public 24 hours a day. A plaza, carved into the center of the towers and integrating the neighboring building as part of its perimeter, forms a new public nexus between the historic district of Kampong Glam and the extension of the city's commercial corridor.

Multiple levels of vertical connectivity give access to large elevated terraces for the hotel and residents, a public observation deck and a sky restaurant atop the office/hotel tower, while establishing a direct connection to the adjacent underground MRT subway station. Vehicular traffic is lifted off the ground to allow uninterrupted pedestrian circulation. Extensive landscape areas at the ground levels, elevated terraces, and roofscapes provide accessible green space equal to 100% of the site area.

The development incorporates environmental strategies through passive and active energy efficient design and naturally ventilated spaces. The building's orientation is optimized to prevailing sun and wind angles, while the concave building massing captures and channels wind flows through and across the site, fostering cool microclimates within the shaded outdoor spaces.

Embracing civic spaces in a symbiotic relationship with each other and thereby transforming the surrounding multivalent urban fabric, the two sculpted towers act as urban space generators.

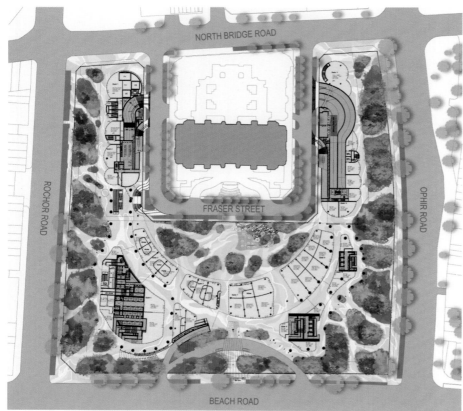

PERMEABLE GROUNDSCAPE

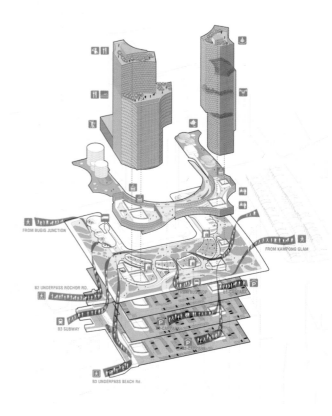

CIVIC NUCLEUS

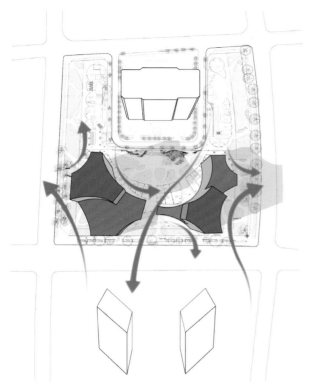

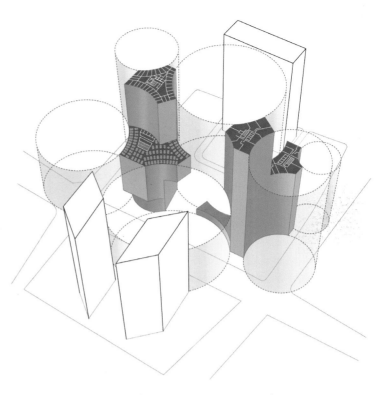

ENVIRONMENTAL FLOWS

BUILDINGS AS URBAN POCHÉ

DUO（双景坊）是马来西亚首相和新加坡总理联合支持，具有历史意义的由国库控股（Khazanah）和淡马锡控股（Temasek）联合开发的项目之一。奥雷·舍人（Ole Scheeren）的设计策略性地呈现出这一富有象征意义的合作关系，并注重与周边环境相融合，使其成为新加坡现代大都会文化的一个新内核。

新加坡一直被评选为亚洲最适宜居住的城市之一。然而，它越来越多地以认为排除社会连系而孤立存在的个人塔为主。规定分区让该项目面临了双重的困境：本项目把场地分割成两个独立的部分，给塔留下巨大的风险，其风险压倒了周围环境和邻近历史悠久的甘榜格南区。

双塔形态的设计是因地制宜，利用一系列循环的圆形来裁切掉建筑体量，消减建筑体量后形成的空间与外部城市空间相互连接，将双塔融入城市文脉之中。

网状的六角形遮阳系统形成天然的外皮附于凹进的外形上，动感十足，部分幕墙形状修长，飞扬延伸至附属的道路。经过对建筑悬臂部分和退线部分的精雕细刻，DUO 就如一对舞伴翩翩起舞，又如太极大师在相互切磋。

基地上设计了一个多孔可渗透的热带景观，消解了大楼直落地面的突兀感。休闲区域和花园作为多个交通枢纽的连接点，建立起一个流动的热带花园和商业环境，供公众二十四小时使用。一个广场被镶嵌在大楼中间，巧妙地利用周边建筑作为它的边界，形成一个新的公共区域，连接 Kampong Glam 的历史区域和城市商业延伸地块。

垂直连通的多个楼层为酒店客户和居民进入大型高架露台提供通道。公共观景台和顶层的餐厅凌驾在办公室与酒店为一体的大楼之上，同时建立了一个直接连接到邻近地铁站的通道。行车交通被转到地面之上，让不息的人流畅通无阻。在一楼宽广的景观区，高架露台和 roofscapes 提供的绿占了场址面积的 100%。

项目同时融合被动节能和积极节能设计实现其环保功能，通过自然通风来造就热带生活的舒适性，最小化能耗并融入自然环境。建筑的朝向已经通过日照角度和风向的优化，凹进形的建筑体量可使得风在基地内自然流动，在遮阳室外空间形成凉爽的小气候。

DUO 两座塔楼坐拥城市空间相互共存，从而改变周围的多元城市结构，成为了城市空间的缔造者。

360° VIEWS

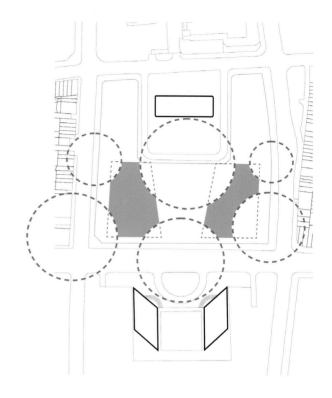

CARVING PUBLIC SPACES

- New City Landmark — Mixed-use Architecture

Architect · amphibianArc
Area · 372,946 m²

Beijing, China
Hongxing Macalline Furniture Beijing Flagship Store

红星美凯龙北京旗舰店

The 372,946 m² flagship store sits on an 89,270 m² site with two stories below ground and seven stories above ground, and 2,362 parking spaces.

Home living has changed drastically in the past 25 years in China. As the largest brand in home furnishings, Hongxing Macalline Furniture not only makes a large selection of home furnishings available to consumers, but also provides new design sensibilities and lifestyles for the modern Chinese home.

The design concept is to create the trip to the flagship store as an experience to "future city" as Hongxing Macalline provides unprecedented shopping experience to the consumer.

The new Beijing Flagship Store represents Hongxing Macalline's new brand image. Architectural elements are used to convey concepts derived from Feng Shui. For instance, the front elevation at the entry is U-shaped, which represents gathering crowd and wealth. Moreover, the overall form of the building resembles a Chinese ingot, which in Feng Shui terms is "treasure-collecting"— an idea very much in line with the Hongxing Macalline's business concept.

Hongxing Macalline has grown in such rapid speed that in 25 years, its stores have spread to 65 cities with more than 80 stores nationwide. The intention for this project is to establish a new brand image which represents Hongxing Macalline's core values with the flagship store in Beijing, and continue to duplicate and spread its brand through its distinct building designs.

In today's fast changing Chinese cities, it's become ever more challenging to differentiate one city image from another. Therefore, creating a distinctive brand identity with variations, which can be implemented to each branch store, such that the differentiated image connects to each local environment, is of the essence.

New City Landmark — Mixed-use Architecture

Variation is achieved with the application of images onto the elevations, and at night LED light transforms and enlivens each building. These images are selected according to the region's culture, unique vegetation, local history, folk art, etc. Once the image is selected, pixilation is applied and translated onto plastic tubes with Hongxing Macalline's logo imprinted onto the face of each tube, and the tubes are then lit by LED lights at night. The pixelated facade is composed of recycled PVC tubes, a material used 150 million tons per year worldwide, and its waste disposition affecting the environment.

This project is commended by 2012 MIPIM Architectural Review Future Project Awards under the Retail and Leisure category.

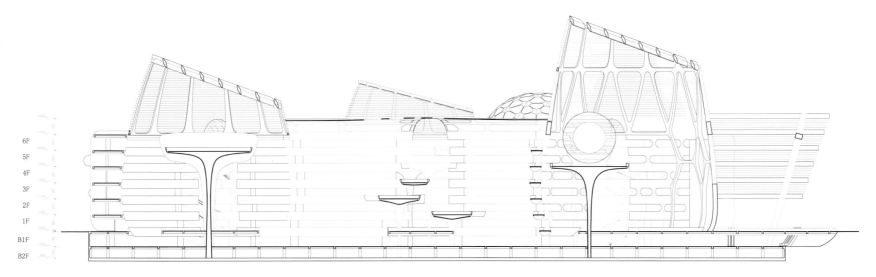

红星美凯龙旗舰店占地面积89,270m²，其中地面七层，地下两层，并设有2,362个停车位。

家居改变生活，25年间，人们对于家居概念的理解发生了巨大的变化。作为中国家居流通业第一品牌，红星美凯龙用家居改变生活，不仅给消费者提供了大量优质的家居选择，更为中国现代家庭展示了一种崭新的设计理念以及生活方式。

红星美凯龙用一种超前的商业手段和引导消费者，在旗舰店的设计中，这种未来商业、未来家居的形象，将为消费者带来前所未有的消费体验和感受。未来城市的意向将贯穿到设计的各个方面，为顾客们开创一种新奇、变化的观感。

北京旗舰店就是红星美凯龙的标志。在南立面上，整个立面呈内陷凹状，是一个寓意汇聚人气、财气的表现手法。在其上的主入口，更是由巨型的突出框架组成，在建筑设计上不仅呈现出大气、开放的入口形象，而且整个建筑，也形似一个元宝。用现代的建筑手法，却依然蕴含了"聚宝"的寓意。

在过去的25年间，红星美凯龙发展迅速，在全国已经拥有80多家门店，覆盖65座城市。设计师的初衷是在北京为

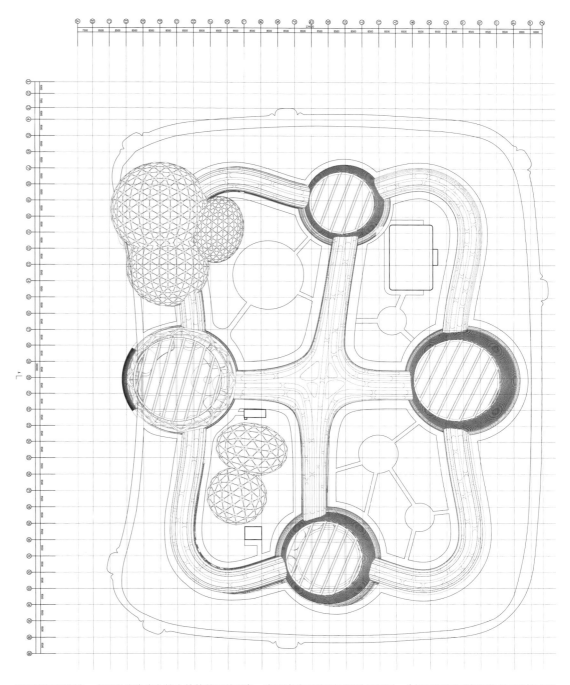

红星美凯龙建造一个可以代表企业核心价值的品牌形象，在不丧失品牌可识别性的同时，为不同的红星店创造不同的地方设计，从而进一步提高其品牌知名度。

在中国的城市化进程中，同质化城市空间问题日益凸显。美凯龙设计师基于相同的设计手法，却会为各个地方城市设计出具有其地理、人文、历史、风貌等特征的建筑立面，为红星创造统一而多变的商业风格。

各地门店的差异性体现在建筑立面图案和夜晚LED灯的变幻组合中。立面图案会依据地方文化，植被，当地历史，民间艺术等进行选择，再用图像像素化处理手法，在建筑立面的塑料管上生成图案，塑料管面上同时印有红星美凯龙的Logo。夜晚，LED灯会将塑料管照亮。立面的塑料管全部采用回收PVC管，以体现设计师对绿色环保建筑的倡导（PVC管全球每年使用量为1.5亿吨，严重影响生态环境）。

该项目获得了2012年MIPIM建筑评论未来奖（零售休闲类）。

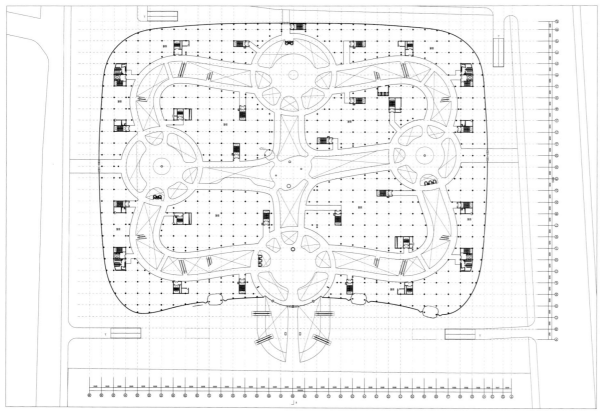

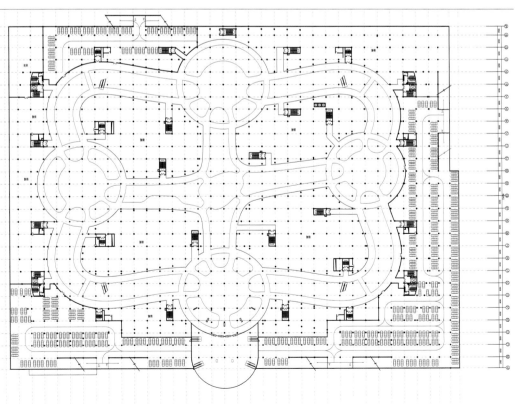

• New City Landmark — Mixed-use Architecture

Located in the Fenghao Boulevard, city's major east-west commercial thoroughfare, the project is planned to be the city's future sub-commercial node, as well as its new fashion center.

This project covers 186 mu, which is about 124, 062 m². The first phase of it covers 20, 569 m², about 31 mu. While the area of the second phase is 103, 493 m², about 155 mu. A holistic design approach was adopted during the project's initial planning phase, to ensure that the first and second development phase can be well-integrated in terms of function and aesthetic. At the same time, a thorough analysis was conducted on the site in order to maximize its commercial value.

As China's ancient capital for six different dynasties, Xi'an is a city lacking in vibe. The project introduced the concept of interior landscape, a new element to the existing cityscape, so that a sustainable and livable environment will always be displayed on-site. Sancai is a great symbol to represent for the Tang dynasty, an era noted for its vitality and colorfulness, it is also an ideal reference for the choice of architectural color. The built project will become an exciting icon for the city of Xi'an.

New City Landmark — Mixed-use Architecture

Architect · amphibianArc
Area · 124,062 m²

Xi'an, China

Xi'an Lijun Mall

西安利君时尚购物中心

The new plan clearly puts forward creating Xi'an as a livable city. Around it, a series of short-term and long-term planning targeted for environmental restoration is made. With the deepening of the understanding about livable ideas, and constantly improving of the construction process livable space is no longer limited to the outdoor space. Green and pleasant interior space will be one direction worth people's efforts to explore the harmonious coexistence of human and nature. Xi 'an is the ancient capital of six dynasties in China, the city color is dignified, and especially in the autumn and winter the whole city is lack of vitality. Project will bring the green home, and add life and color for the existing dignified and grey Xi 'an, making sure that it's like spring all the year round in the project area.

本项目位于陕西省西安市市东西主干道丰镐路，规划为城市未来商业次中心以及新的时尚中心。

项目区占地124 062 m²。其中，一期占地面积20 569 m²，二期占地面积103 493 m²。项目在最初规划上采用了全盘考量，使得两期开发在功能及整体建筑效果上可以得到很好的衔接，设计理念可以贯穿整个项目。同时，项目在最初设计阶段对项目区位进行了全面研究，各功能板块依据分析结果进行布置使其商业价值得到最大化。

西安作为唐代的中心，其时盛行的唐三彩以其色泽绚丽，富有生活气息而闻名，将是本项目最佳的颜色参照。唐三彩以色泽艳丽和富有生活气息而著称，以该颜色作为建筑主色将为西安市增添一道亮丽的城市色彩。

新一轮规划明确提出将西安打造为宜居城市。并围绕该定位制定了一系列以环境复原为

目标的近期及远景规划。随着人们对宜居概念理解的深入，及建筑工艺的不断提高。宜居已不仅仅局限于室外空间。绿色宜人的室内空间也将是人们探索人类与自然和谐共生的努力方向之一。西安作为中国六朝古都，城市颜色凝重，尤其在冬天秋天，整个城市更是缺少生机。项目将引绿入室，为现有凝重并暗灰的西安增添生气及色彩，确保项目区四季如春。

• New City Landmark — Mixed-use Architecture

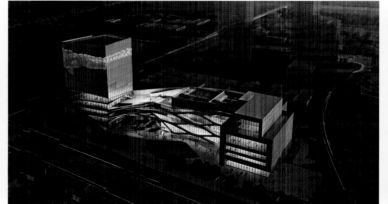

New City Landmark — Mixed-use Architecture

Architect · BLUA
Design Team · Bin Lu, Shou Zheren, Ge Liangjun, Chen Jian, Qian Haiping, Ma Li, Jiang Tianjiao, Zhang Hong, Zheng Jue, Xia Tingting, Tao Jiangjie, Shen Qing. David Stavros
Client · Hangzhou Sports Development Group

Hangzhou, China

Hangzhou Civic Sports Center

杭州文娱体育中心

The site for this project is located at the west side of Hangzhou new city, 2.6 km to the CBD center. The facility has a total built area of 15,500 m². Above ground floor area is 46,700 m². The southwest corner is the entrance for the planned Metro Line and around the site are high-end residential districts. The north side is bounded by a primary school. Adjacent to Qian Jiang River, the site is a connector between the natural and urban life of the city, making it a perfect location for a sports complex. The project is based on creating an urban plaza while also creating an icon without large-scale commercial facilities and with distribution of leisure spaces. With the convenient transportation, the sports center radiates across the whole city.

BLUA create a new synthetic ground for sports activities while also an iconic figure against the relaxed background of the sports landscape. This building is a symbol of the new status of Hangzhou as one of the six ancient capital and international city. From a massing view, the designers design the podium and 42m tower in a stacked way in order to contrast with the complexity of the 70m tower. The roof is interconnected to make a continuous, differentiated sports landscape with cascading sports fields and pathways. This landscape features basketball courts, volleyball courts, badminton, and tennis, at various levels. The steel frame of the podium becomes the major structural piece for the curtain wall but also houses the circulation element for the facade. The sunken plaza becomes a big open space which has been designed as a microclimate ecological environment facing the school. It also can be entered from the civic square on the southeast. The set back of the south side creates a rectangular civic square. The civic square and the green belt integrate into a 6,000 m² long main entrance space. The civic square distributes the flow of people pressure from the main eutrance and the Metro Line and is covered with landscaping.

The sunken plaza is 1.5 m underground, which connects with the entry square by folding shaped steps. The garden plaza is shaded via a cantilevering roof, and a series of pools underneath allow for evaporative cooling to occur, while also interacting with the swimming pool behind the curtain wall, creating a lively water environment. It is spatially defined on all sides by topography, plantings, shelters and the plaza roof, creating a highly specific and integrated environment and small microclimate. The plaza roof and canopies offer shade, while plant-life and water features cool the air underneath based on principals of natural convection and evaporative cooling, creating a thermally regulated environment. This space is a lush and tropical atmosphere including public amenities such as gift shops, bookstores, and cafes, allowing people sanctuary from the stresses of city life.

According to the different types of space, they divided into two parts: "active" and "inactive". Sports activities are located on the lower part of the tower and the podium, in clear groups for ease of orientation piled into a single piece of "active" area. The top of the tower is planned as a relatively "inactive" area with offices, accessories and the VIP Club.

The main tower and podium are wrapped in three folding pieces. The twisted-folding shape has a dynamic trend of an upward spiral, which makes the tower an increasingly upright illusion higher than 70 m. The trend of folding shape determines the density of the skin texture.

The design features faceted crystalline geometry embedded with crystal patterns in 3D cold lamination film, ETFE membrane, and honeycomb glass structure. They are used to heighten the sense of irresolution between flatness and depth as well as correlate graphic/pattern effects with mass inflections. The design also features massive membrane bubble "windows" oriented to allow views out to all angles of the city. The perforated stainless steel panel on the wall strengthened the crystal shape, as an attraction landmark for the city.

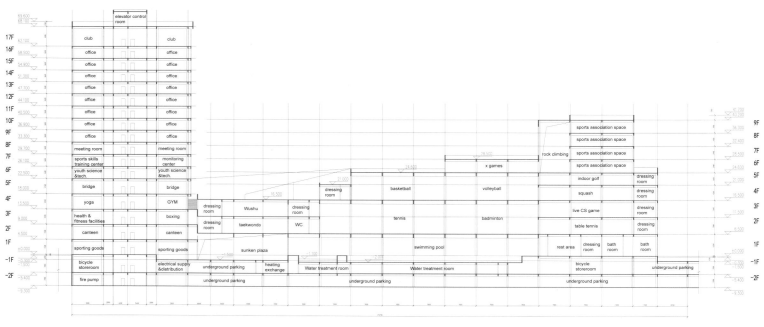

1-1 剖面图
1-1 Section

该项目位于杭州新城区西侧，离 CBD 中心 2.6km。该设施的总建筑面积为 155 000 000 m²。地上建筑面积为 46 700m²。西南角是规划中的地铁线入口，附近是高端住宅小区。北边以一小学为界，毗邻钱塘江，该处作为连接自然和城市的纽带，使其成为建设体育中心的绝佳地点。项目基于建造一座城市广场，同时开创不含大型商场设施和休闲空间的先河。因为交通便利，体育中心可以辐射整个城市。

BLUA 为体育运动创作了新型合成地面的同时也为宽松的运动背景增添了一个标志性形象。作为六大古都和国际城市的杭州城，该建筑成为杭州新状态的象征。设计采用堆叠方式设计领奖台和 42m 高的大楼主要是为了和 70m 高的大楼形成鲜明的对比。屋顶是内部连通的，营造一种连续的和阶梯式运动场和跑道，有着不同的运动景观。这些景观在不同层次上为篮球场，排球场，羽毛球场和网球场增色不少。领奖台上的钢架成为幕墙的主要构件，同时也支撑着表面的循环元素。下沉广场变成一个大的开放空间已被设计成一个面对学校的小气候生态环境，同时它也可以从东北边的市民广场进去。在南侧建造一座矩形市民广场。市民广场和绿化带一起组合成 6 000m² 的入口主体。市民广场可以有效缓解来自地铁的人流压力，同时还覆盖着自然植被。

下沉广场于地下 1.5m，通过折叠型台阶与入口广场相连。通过悬挑屋顶形成花园广场阴影区域，一些底下水池不停地水分蒸发冷却。在幕墙后面还设有游泳池，营造热闹的水上环境。从地势上、植被上、候车亭、广场屋顶等每一个方面都在进行空间定义，造就非常具体的集成化环境构造的小气候环境。广场屋顶和天棚提供了阴凉，底下的水分通过自然对流和蒸发冷却调节空气的温度，创建一种热调节环境。该空间充满郁郁葱葱的热带氛围，包括公共设施，如礼品店、书店和咖啡馆，成为人们逃避城

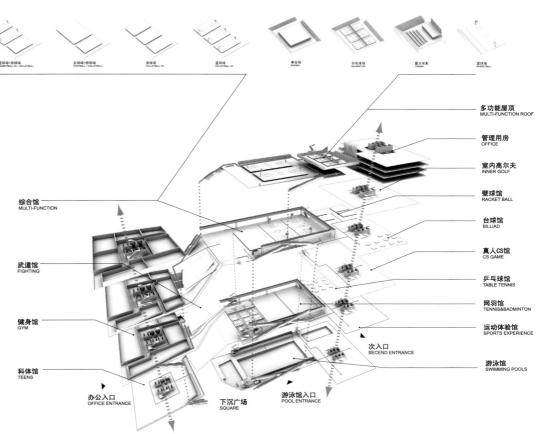

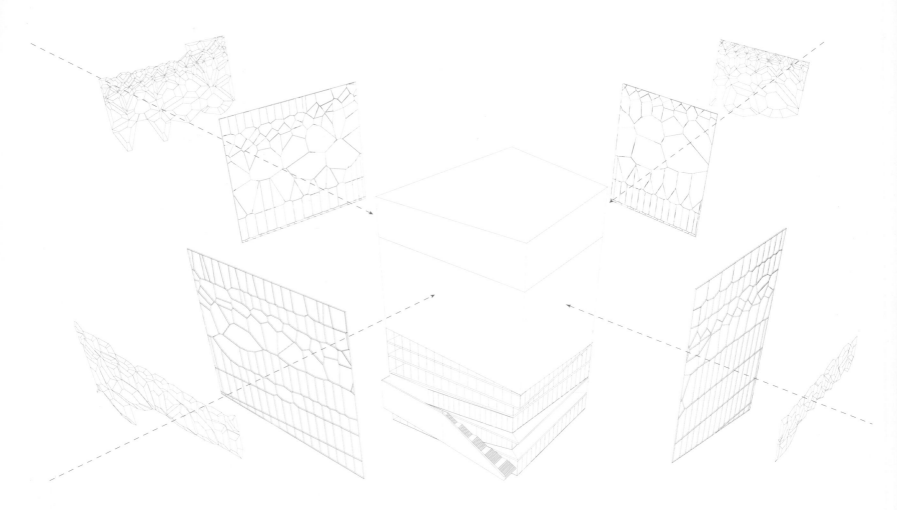

市生活压力的避难所。

根据不同类型的空间，设计师将其分为活动区和非活动区两部分。体育活动区位于大厦和领奖台的底部，毫无疑问归类为活动区。大厦顶部计划作为一个相对"不活跃"的办公区，配件区及VIP俱乐部。

主楼和领奖台由三个折叠构件所覆盖。扭曲的折叠形状有一种动态的螺旋上升的趋势，这使得对大厦直立高度有超过70m的错觉。折叠形状的走向决定了建筑群表面的密度。

该设计采用几何嵌有水晶图案的3D冷裱膜，ETFE膜，和玻璃蜂窝结构。它们用来加强平整度和深度之间的难以臆断的感觉。该设计采用大量的膜泡窗口为主导，以欣赏到城市的各个角落。墙面上的穿孔不锈钢板强化了晶体形态，成为吸引观众的城市地标。

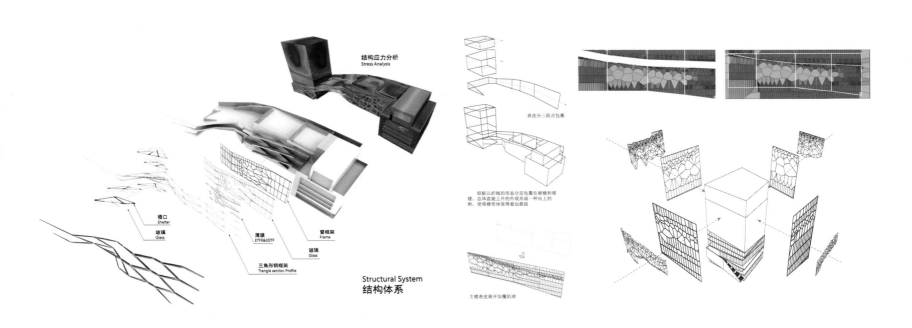

Structural System
结构体系

• New City Landmark — Mixed-use Architecture

Beirut, Lebanon

Matn Point

Matn Point

Architect · Design International
Project Principals · JDavide Padoa (CEO), Lucio Guerra (Managing Director), Paul Mollé (Founding Partner) **Area** · 70,800 m²
Program · Retail, Entertainment, Restaurant, Hotel

New City Landmark — Mixed-use Architecture

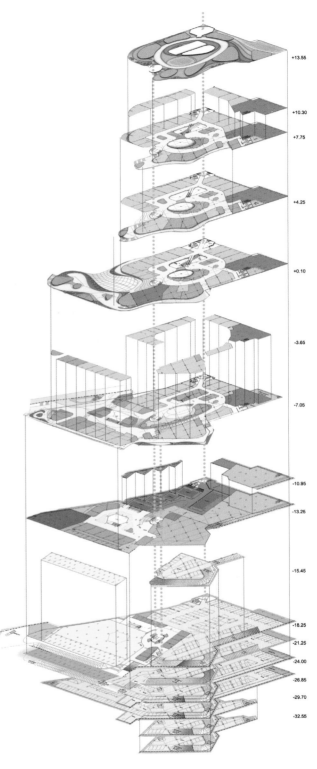

In 2012, developer SIDCOM teamed up with international retail architect Davide Padoa of Design International, to create a unique shopping and entertainment destination for the people of Beirut, the wider region and international visitors. Combining a unique location, stunning architecture and the best retail, dining and entertainment offers with the most breathtaking view in the whole of Lebanon, MatnPoint will be the first designer resort in Lebanon.

MatnPoint offers a unique merchandise mix across 6 levels with over 150 shops, more than 5 anchor stores and a large number of entertainment and leisure facilities. The project is divided into a neighbourhood community centre on the lower floors with hypermarket, kindergarden, family entertainment centre and a Spa boutique hotel. The upper floors, which cater for the wider Beirut population as well as for the large number of tourist coming to the area, contain entertainment facilities and a cinema complex, the first off price department store in Lebanon, temporary & pop up stores with new offers every season, a large selection of fine dining restaurants as well as bars and a nightclub on the roof terrace.

The architecture takes inspiration and contextualizes the surrounding, by creating curved lines that shape the building upwards, which seem like a natural extension of the hilltop. The terraces on the various floors harmoniously integrate into this architectural language, while offering stunning views of the area. The view is most spectacular from the roof terrace, where guests can see across an infinity pool and over Beirut and the sea in the distance, while one of the largest waterfalls ever designed in a mall reaches down from the roof terrace several levels into the project.

2012年，开发商SIDCOM联手设计国际零售建筑的，Design International 公司的设计师Davide Padoa，为贝鲁特人以及国际游客建造了一个独特的购物和娱乐中心。绝佳的建筑与最好的零售、餐饮和娱乐结合，独一无二的位置，将打造黎巴嫩最为惊人的建筑景观，MatnPoint 将成为黎巴嫩的第一个设计师胜地。

MatnPoint 将是一个独一无二的商业综合体，六层的场地包含150家店铺、5家主力店和大量的娱乐和休闲设施。项目分为一个邻里社区中心和一个水疗精品酒店，邻里社区中心在较低层，包括超市、幼儿园、家庭娱乐中心。较高层为范围更宽广的贝鲁特人和大量的游客提供服务，包含娱乐设施、电影院、黎巴嫩的首家特价百货商店、季节性临时商店、大量精选高级餐厅以及屋顶露台的酒吧和夜总会。

通过塑造建筑体曲线，将建筑融入周边环境，像一个山顶的自然延伸。不同楼层的露台和谐地融入到建筑语言中，为该区域提供了魅力景致。站在屋顶露台往下看，景观相当壮观，客人可以看到整个游泳池和远处的大海；在商场内设计的一个最大的瀑布，沿屋顶露台而下，经过几层，进入项目内。

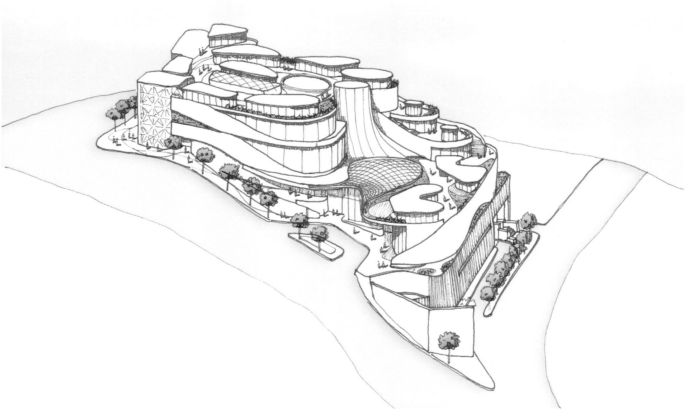

- New City Landmark — Mixed-use Architecture

New City Landmark — Mixed-use Architecture

Nanjing, China

Nanjing Studio 14

南京 14 工作室

Architect · HWCD
Design Team · JHuang Shuiying, Qu Xishan, Chen Yunjin, Fengruoqian, Matyas Simonyi, David Valent, Adon Buckley, Henry Sykes, Octavio Gonzales
Client · Studio 14 Research Institute of Electronic Technology Group Corporation
Program · Retail, Office, Residential, Entertainment

International architecture practice, HWCD has developed a series of mixed-use towers located at Yangtze River Delta. The team of The Nanjing Studio 14 submitted the project as part of a competition to establish an important hub in the center of Nanjing, capital of the Jiangsu province.

As a design firm they specialize in urban planning, architecture and interior. Every project undertaken incorporates the local cultures into the design process and has an emphasis on research development. In order to have a firmly rooted cultural and environmental context HWCD architects come from varying international and domestic backgrounds and work together to produce original and exclusive designs under a modern architecture paradigm.

The design team drew inspiration from the area, looking at the language of traditional local paintings depicting the landscape, while still working within what is a very modern and expanding metropolis. Nanjing was conceived as a "high-tech base of innovation".

The use of a variety of new technologies, such as rainwater harvesting, solar collectors, and geo heat pumps are implemented for environmental protection and energy saving purposes. Studio 14 entry attempts to combine the traditional, natural and modern in a harmonious addition to the skyline, thus far creating a new landmark for the city.

The development houses a multitude of uses, office facilities, mall, leisure facilities, sports and public space, as well as a research component. The solution was a series of towers, two ultra-high rise and one high-rise, occupying a base land area of 109,100 m^2, connecting these are indoor and outdoor public space, landscape and vegetated trellises which climb over the towers striking forms.

The shape of the towers evokes the traditional mountain paintings; their tapering forms recalling the angular "faces" of the mountains. Around this, systems of landscape, water and circulation flow, inspired by the local hanging gardens often depicted in the local artwork. The towers use a limited palate of materials, predominantly brick and glass curtain walling, the grey-blue brick inspired by the ancient city walls of China and the sense of scale they inspire. Local planting and materials help fully integrate this new development into the area.

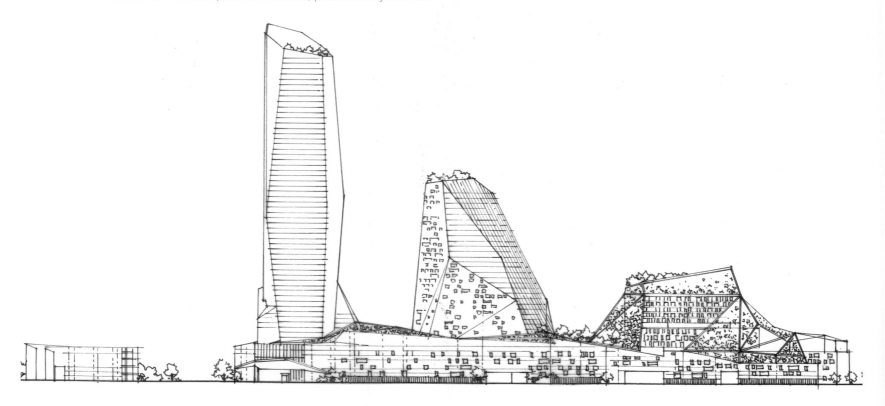

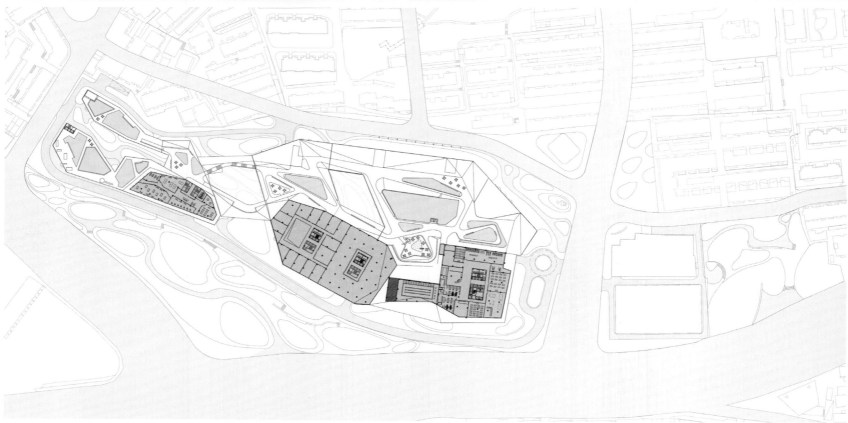

国际建筑机构 HWCD 开发了一系列扬子江江畔的综合体。设计团队将南京 14 工作室作为南京核心地带的一个重要中心来打造。

作为一个设计事务所，他们专门从事城市规划、建筑和室内设计。在每一个他们设计项目中，他们都将当地文化融入到设计中，并强调研究开发。为了让该项目拥有植根的文化和环境背景，来自不同国家的建筑师一起合作，在一种当代建筑模式下创造出原创的、独一无二的设计。

设计团队从建筑所在区域获得灵感，他们查看当地传统景观绘画的语言，将它们运用到南京这个相当现代而且不断扩大的大都市的建筑中。

利用各种新技术如雨水收集、太阳能集热器、地热泵，实现环境保护和节能。14 工作室试

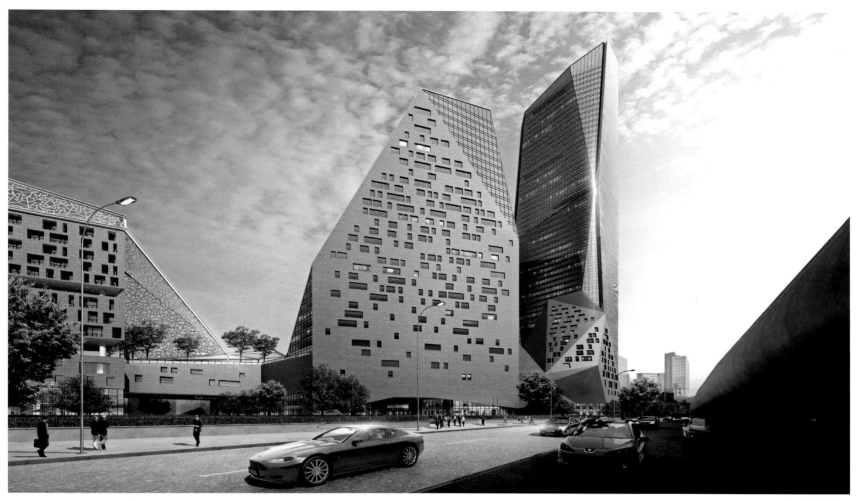

图以和谐的方式结合传统、自然和现代，建造高耸入云的建筑，从而将其创造成一个新的城市地标。

该综合体要求具有各种功能空间，如办公设施、商场、休闲设施、体育、公共空间以及研究所。解决方案是一个系列的大楼即两栋超高层和一栋高层，占地面积为109 100m²，连接它们的是室内室外公共空间、景观和以引人注目的方式爬上大楼的植被盆。

大楼的形状让人联想起传统的山水画，它们的圆锥形式仿佛山脉的峰角。灵感来自当地空中花园的景观、水和环流体系常常出现在当地的艺术画中。大楼采用有限美感的材料，主要是砖和玻璃幕墙，其中灰蓝色的砖是受中国古城墙及其规模感的影响。当地植被和材料帮助这个新开发区完全融入到该地区。

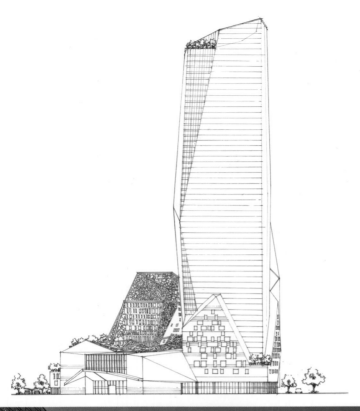

- New City Landmark — Mixed-use Architecture

Architect · HWCD
Client · SRE
Program · Retail, Residential, Restaurant

Shanghai, China

SRE Riverside Apartment Complex

SRE 河滨公寓

New City Landmark — Mixed-use Architecture

This project is located in East Longhua Road, Luwan district, Shanghai. Luwan district is one of the areas in Shanghai that exemplifies the blending oriental and occidental culture. With the successful holding of the 2010 Expo in Shanghai, Luwan area becomes very active.

Because the site doesn't have visual connection to the Huangpu River and it is beset on all sides by high-rise buildings, the designers created an internal landscape capable of attracting people. Car traffic and parking were located underground in order to maximize green areas and pedestrian roads. The complex offers a wide variety of activities for residents (open swimming pool, shopping mall, retail shops and restaurants).

The buildings draw their inspiration from two main concepts: First, inspired by the boats and yachts moving along Huangpu River. Second, based on natural shapes that generate generous double height terraces.

Indoor and outdoor spaces are strongly connected and curve shape of the buildings makes a softer impact on the site than the surrounding buildings.

该项目位于上海卢湾区龙华路东。卢湾区是上海的一个区，它充分体现了东西方文化的交融。由于成功举办 2010 年上海世博会，卢湾区变得非常活跃。

由于场址受周边的高层建筑遮挡，在这不能直接看到黄浦江，设计师创建了能够吸引人的一个内部景观。汽车交通和停车场都位于地下层，以使得绿色区域和人行道路最大化。综合体为居民提供了各种各样的活动区域（开放游泳池、购物中心、零售商店和餐馆）。

建筑师的灵感来自两个主要理念：
首先，灵感来自于沿着黄浦江的船、游艇。
其次，基于生成宽阔双高露台的自然物形态。
室内和室外空间强烈联系起来，建筑的曲线形状使得该建筑产生一个比周围建筑较为柔和的印象。

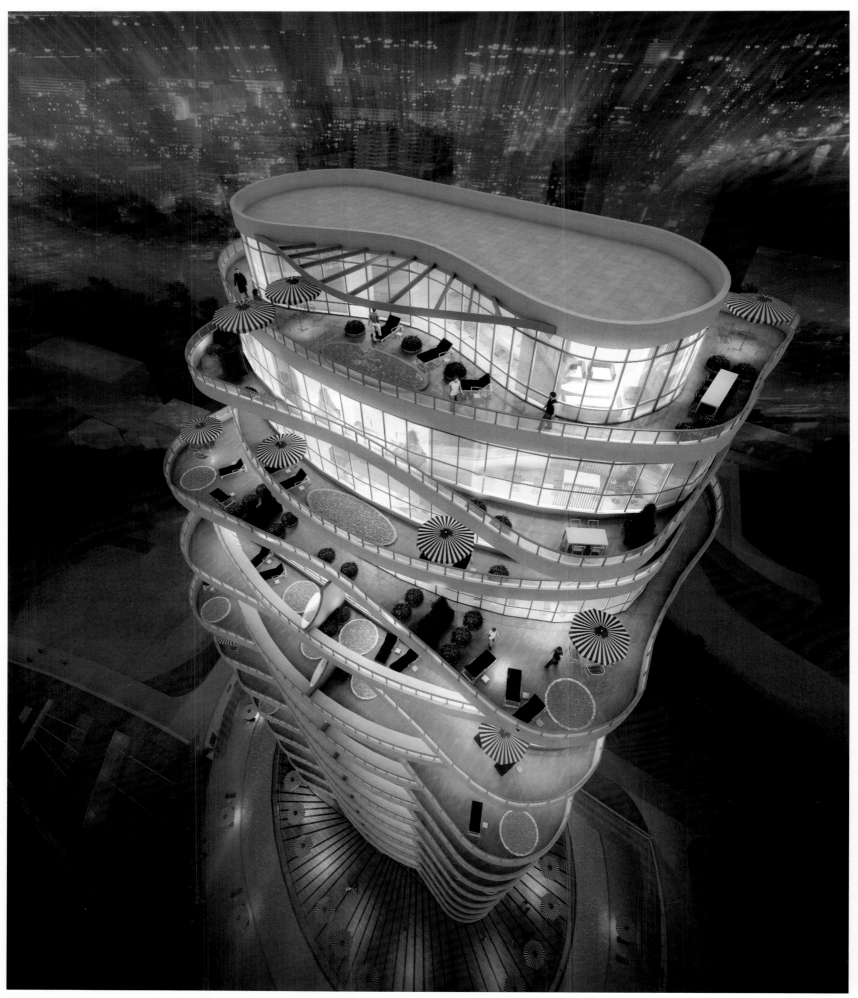

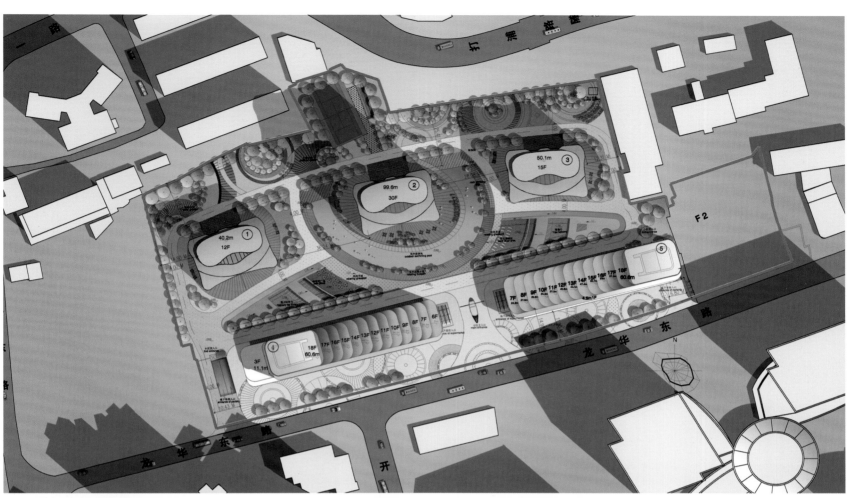

New City Landmark — Mixed-use Architecture

Architect · KaziaLi Design Collaborative
Designer · Clay Vogel, AIA
Client · Hunan ARTEL Guangsha Real Estate Development Co., Ltd.
Area · 85,980 m^2
Program: Office, Retail, Entertainment, Parking

Changsha, China

Changsha Shimao Plaza

长沙世茂广场

The Shimao Plaza is located in the central business district of Changsha, Hunan province. The design concept is based on the natural landscape of Hunan which is lush with mountains and rivers. The towers take the form of stones which have been worn and polished by the rushing current over the years. The larger office tower stands at 400 m tall and smaller tower at 120 m. They both sit on a stone slab which becomes the commercial area of the project. The design also features a 33 m high waterfall, acting as cooling element, as well as an anchor attracting people to the site.

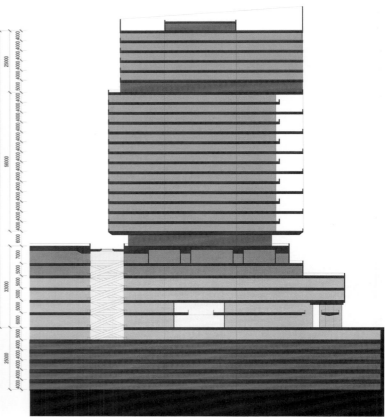
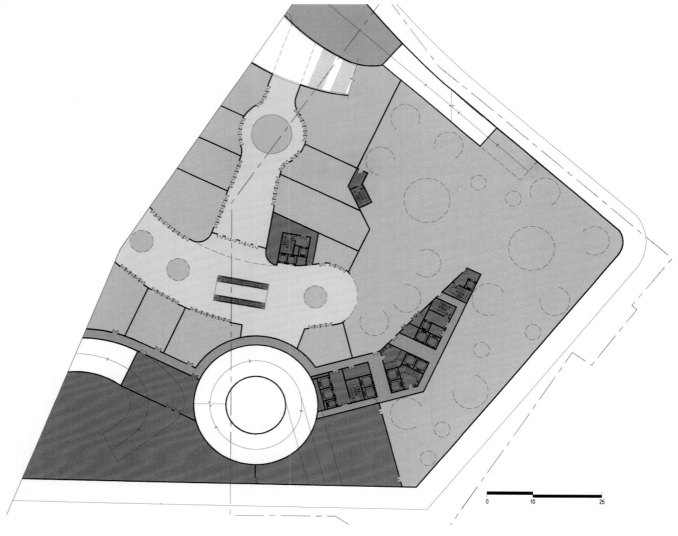

长沙世茂广场项目位于长沙市核心区域的长沙芙蓉中央商务区（CBD），正处于中央商务区的东西轴线五一大道与南北轴线芙蓉路的交汇的"金十字"地带的中心点位置。项目北向为五一大道，东侧为芙蓉中路，沿地块边有链接五一大道与芙蓉中路的建湘路。高的办公大楼有400m高，矮的办公楼有120m高。两座大楼都建在石材基座上，此基座构成了项目的商业部分。此设计的一个特点是有一个33m高的水景，既是消暑的一种设计，也成了吸引人群的一个亮点。

• New City Landmark — Mixed-use Architecture

New City Landmark — Mixed-use Architecture

Architect · LAVA
Design Team · Chris Bosse, Tobias Wallisser, Alexander Rieck, Jarrod Lamshed, Angelo Ungarelli, Vivienne Ni, Paul Bart, Giulia Conti, Manuel Caicoya, Sandra Pamplona, Guido Rivai, Alessandra Moschella
Program: Hotel, Residential, Retail

Shenzhen, China

Shenzhen Jungle Plaza

深圳森林广场

This self-sustained project comprises a 5 star luxury hotel, residential, retail, urban village and parklands. The design embraces the local terrain of mountains, tropical flora and water systems by integrating an artificial water mass.

Subtropical flora, dense jungle foliage and rice terraces create a lush environment. The existing greenery on site is recycled and integrated into the building design. Full hanging foliages flow from the exterior of the buildings, integrating nature and the built environment.

Sustainable features include water collection that feeds the gardens and iintelligent solar reactive louvres that respond to the sun's direction and keep the building interiors cool.

Integrated transport systems including cable cars connect each of the buildings, making the site car-free. At ground level connectivity is facilitated through timber footbridges and paddleboats on the lake.

A "see-through" canal, its shape imitating a drop of water, gives fascinating glimpses into an underwater world and brings light to underground levels. Solar powered "Evolution" light poles create ambient light.

The organisation of the 32-floor hotel is based on its shape: podium floors house public activities, whilst the tower is dedicated to luxury suites. Room services are moved to the edges and voronoid shaped surfaces create a fluid room space with no walls.

Three residential towers taper in height towards the edges of the site to embrace and protect the landscape. The orientation is north south to allow natural cross ventilation, providing nine hours sun. Terraces from the podium continue up every floor, enriching the apartments with wide balconies.

A lift in the middle of an indoor aquarium takes shoppers at one retail mall up five levels of fashion to a rooftop "Sunset Bar", whilst another mall features general retail with an outdoor cinema.

LAVA integrates nature and the built environment, utilising the geometries in nature to create beautiful and efficient design solutions.

项目设计上遵循能源完全自我维持发展的宗旨，建筑中将包括一家五星级豪华酒店，SOHO住宅，高档时装零售，城中村和绿地。设计充分利用深圳的热带气候影响，通过集成一个人工的水资源循环利用系统，设计融合当地山区的峡谷和自然水系。

亚热带植物，茂密的丛林树叶和梯田，创造出一个郁郁葱葱的环境。设计将回收利用融入到建筑当中。建筑物的外墙全部挂满树叶枝条，从而实现自然和建筑环境的统一。

可持续发展的功能包括雨水收集、水资源采集的花园和太阳能百叶窗，它能随太阳的方向调节反射角度，保持建筑物的室内凉爽。

综合交通系统包括连接各建筑物的缆车，使整个项目地域之内实现无车交通。在地面，设计通过木制的人行天桥和湖上的摆渡船实现相互连接，达到交通的便利。

一个透明的管道，其形状模仿一滴水，让人隐约看到一个迷人的水下世界，并给地下层

带来光线。运用太阳能的路灯杆产生照亮四周的光线。

32层的酒店根据它的形状而分布：裙楼用于公共活动，而塔楼致力于奢华套房。服务间移动到边缘，鸟喙状的表面创造了一个流动的空间，没有幕墙。

靠向场址边缘的住宅大楼在高处逐渐变尖，以保护景观。住宅方向是南北向的，拥有自然对流通风以及9小时的太阳光照。从平地开始，露台存在于每一层，宽阔的阳台丰富了公寓景观。

电梯在一个室内水族馆的中间，消费者经过一个零售商场的五层时装区能到达屋顶的"日落酒吧"，而另一个商场的特色是商场与露天电影院。

LAVA将自然与建筑环境融为一体，利用自然界的几何形状创造美丽而有效的设计方案。

• New City Landmark — Mixed-use Architecture

Triangle de Gonesse, France

Europa City

欧罗巴市

Architect · MANUELLE GAUTRAND ARCHITECTURE
Design Team · Olivier Evrard, project manager
Collaborators · TESS, TRANSSOLAR, BASE, TRANSITEC, MICHEL FORGUE
Partner In Charge · Bjarke Ingels, Andreas Klok Pedersen
Project Leader · Jacob Sand, Joao Albuquerque
Client · Groupe Auchan
Area · 800,000 m²
Program · Retail, Hotel, Convention

New City Landmark — Mixed-use Architecture

Europa City is an ambitious project based on the idea of creating an absolute must and stop place, at the national and international scale. The objective is to show and explain the European features, country by country, with a project mixing retail, leisure and cultural activities. Each European country is shown through a specific location.

Dedicated to the French but also to all the tourists who come to visit Europe, the goal is to give a first vision of Europe, before continuing the trip.

Project is taking place in a complex site, along the motorway between the Charles de Gaule Airport in Roissy and Paris. With its very big program, around 600, 000m², the project is supposed to upgrade site and surroundings. The project will certainly be the beginning of a complete improvement of this north part of Paris linked to Roissy.

The architecture develops itself in a "star shape", corresponding at the same time to several goals: to the program needs and to the context needs.

Each branch elongates itself into the landscape, allowing surroundings to be closer and mixed with indoors and outdoors programs.

Roof is a sort of technological skin, mixing in a sensible way a lot of sustainable goals: water is collected and recycled, solar energy is created with more than 20, 000m² of photovoltaic cells, greenery protects and isolates indoors spaces. Some public spaces are created on the roof too, emphasizing beautiful views on the Parisian plain.

欧罗巴市是建筑师雄心勃勃的建筑项目，建筑师想以全国性的、国际的规模将它创建成一个人们"必须、绝对"会停留的地方。项目的目标是展示和诠释欧洲各个国家的特色，包括零售、娱乐和文化设施。每个欧洲国家都通过一个具体的地方展现出来。

这个项目主要面向法国人，但也面向所有来欧洲的游客，在游客开始旅行前展现给他们欧洲的初步印象。

项目在一个复杂的场址上，沿着鲁瓦西和巴黎的戴高乐机场之间的高速公路。由于它近

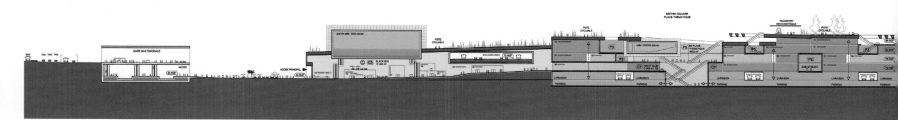

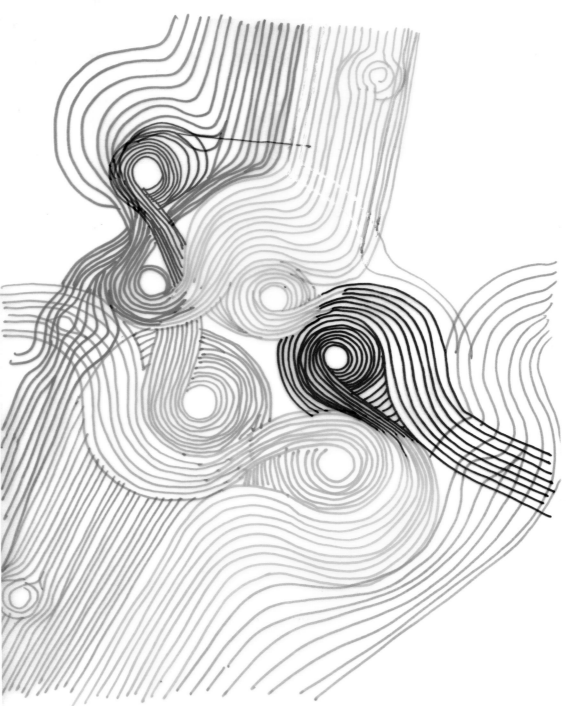

20 000 ㎡，规模庞大，被预测为将提升场址和周边环境的设施和经济。建筑师将其开发成"星状"，同时响应几个目标：满足项目和环境的需求。每个分支延长到景观中，让环境更为亲近，同时融合室内外项目。

屋顶是一种技术性的表皮，以一种合理的方式融合许多可持续目标：水被收集和回收，超过 20 000 ㎡的光伏电池创造太阳能，绿色植物保护和隔离室内空间。在屋顶上创建了一些公共空间，也强调巴黎平原的美丽风景。

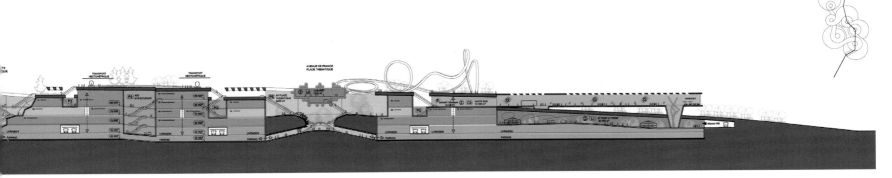

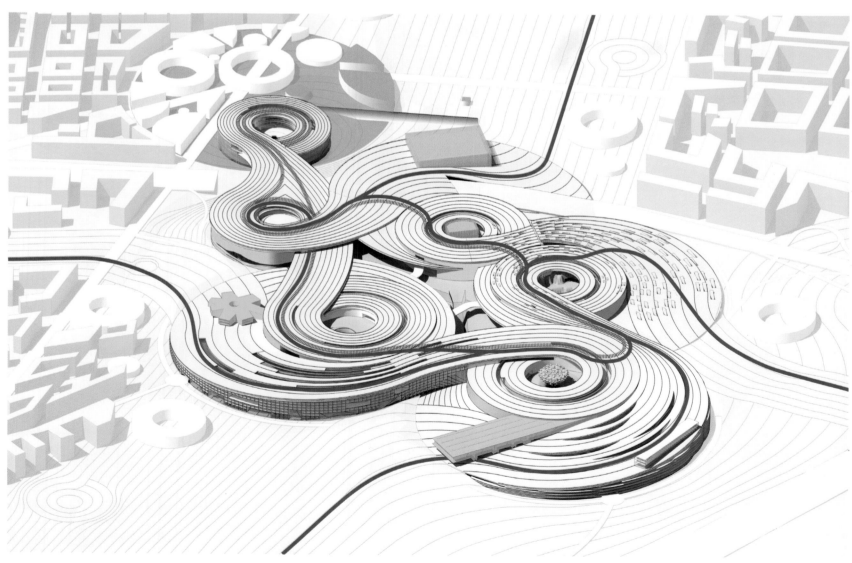

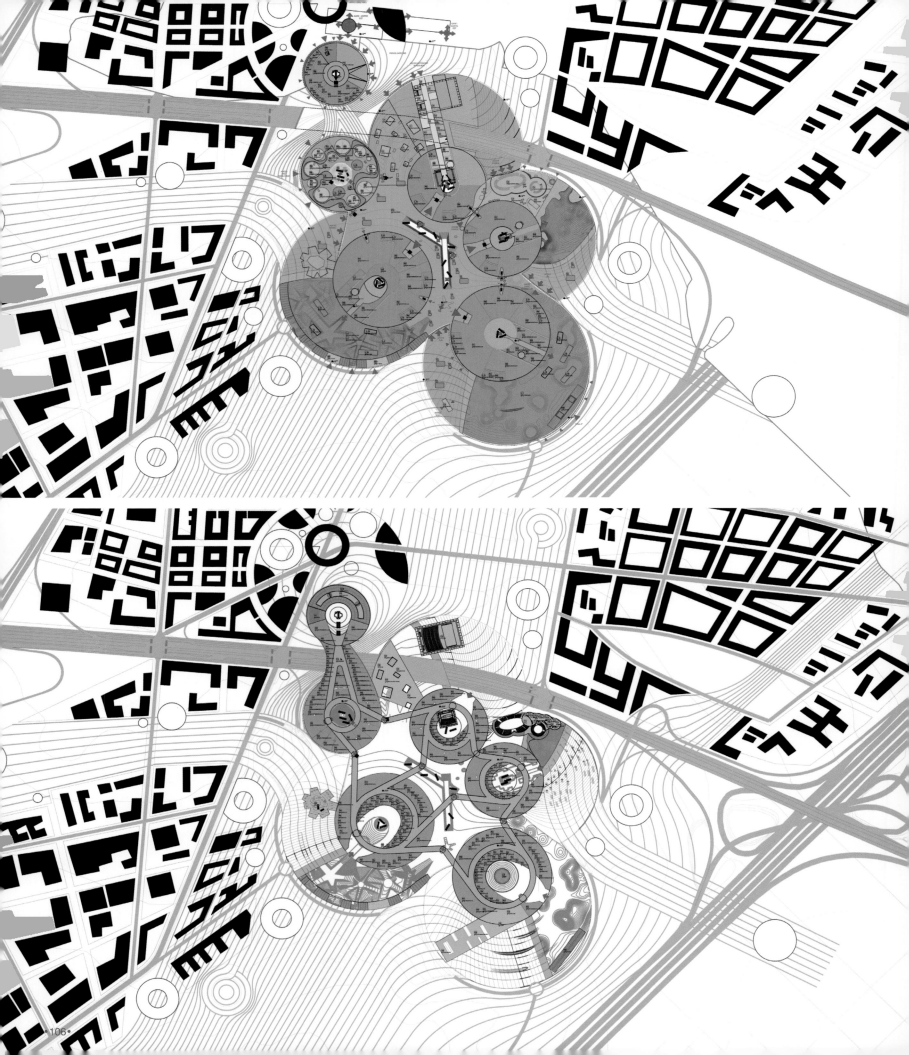

- New City Landmark — Mixed-use Architecture

New City Landmark — Mixed-use Architecture

Architects	SDA|Synthesis Design + Architecture
Design Team	Synthesis Design + Architecture (Design Architect), Shenzhen General Architectural Design Institute, (Executive Architect, Structural Engineer, MEP), OneView CG (Visualization)
Client	Hong Kong Wuzhou International Group Co., Ltd.
Area	180,000 m²
Program	Mixed-use Office, Hotel, Retail, Entertainment & Lifestyle Development

Shanghai, China

Shanghai Wuzhou International Plaza

上海五洲国际广场

Situated along Huatai Road in the third ring of the urban metropolis of Shanghai, the Shanghai Wuzhou International Plaza embodies the energy and vibrancy of the cities distinct urban environment. Inspired by traditional Chinese concepts of Yin and Yang, the "Urban Canyon" is organized as two nested rock-like volumes which have been broken apart to reveal a flowing canyon condition which connects the project to the urban fabric of the city. The northern block features an enclosed 4 storey luxury retail shopping podium anchored by the corporate headquarters of developer Hong Kong Wuzhou International Group and a 5-star hotel tower. The southern block is composed of a 4-level retail, lifestyle and entertainment complex anchored by two office towers. The fluid canyon condition connects the two entry plazas of the site with a "river" of free-standing detached retail units with a network of connective sky bridges, while simultaneously curating a series of framed views within the site. A series of green space "islands" are distributed within the river to provide natural shading and to soften the urban condition. At the mouth of each canyon is a landscaped entry plaza framed by the portal created by its respective towers. Integrated landscaping, furnishings, and lighting within the plaza hard-scape are arranged in pulse like formations which stimulate and encourage visual and pedestrian activity.

The dynamic patterning of the plaza is further expressed in the striated articulations that define the pattern of the cladding. This pattern embodies the pulses of activity and urban energy of the city to merge façade with roof and podium with tower, which is conceptualized as the river that has carved the canyon. The façade is to be clad with RHEINZINK standing-seam titanium zinc panels, while the roof system utilizes RHEINZINK double standing-seam titanium zinc panels. The roof area is equipped with interior gutters at its lowest points and is covered with perforated standing-seam profiles to protect it from soiling. The building material needs no maintenance because of the material's patina, which develops during the course of natural weathering and protects it from corrosion. The patina is a layer of zinc carbonate, which regenerates itself.

本案位于上海三环线上的华泰路，将上海这座中国经济中心城市独有的都市生机与活力尽显无遗。设计灵感源于中国传统文化"阴阳太极"的概念，"都市峡谷"犹如两座阴阳镶嵌的巨石，中间形成一道流动的峡谷地带。北区块是由一座5星级酒店高楼以及开发商香港五洲国际集团的公司总部与四层裙楼组成的豪华零售购物商场。南区块由两幢办公高楼及四层裙楼的综合楼（零售、生活、娱乐）组成。由南北两个建筑体块"阴阳"互生而形成的中心峡谷"河流"地带，则是由一幢幢独立的零售建筑体组成，并由天桥将它们有机连接在一起，形成有序

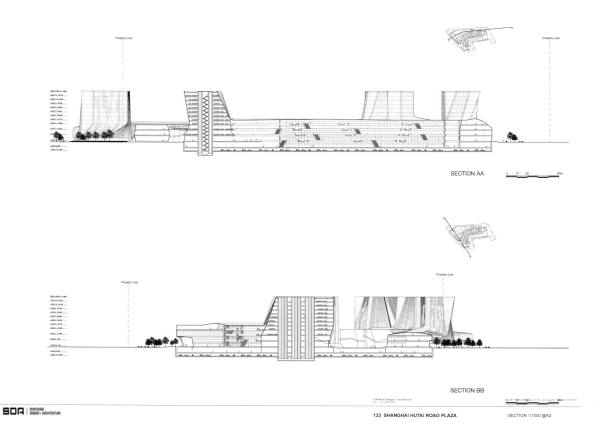

的建筑整体。在这条"河流"地带植入了大大小小的绿色景观，营造自然的氛围以弱化坚硬的建筑群。峡谷的两端入口则是由标志性的塔楼与景观广场组成。统一的景观规划与照明设计使这座"都市峡谷"更具视觉欣赏、充满活力与动感。

包层上的条纹进一步呈现了广场的动态模式。这种模式展现了城市的运动脉搏与能量，宛如河流冲刷而过，形成峡谷。这座由"阴阳太极"概念互生而形成的建筑体量，曲线优美，动感流畅，建筑的表皮与顶面都是由德国RHEINZINK"莱茵辛克"钛锌板构成，表面经过特殊碳化处理，可自我修复具有高强抗腐蚀能力，无需维护。

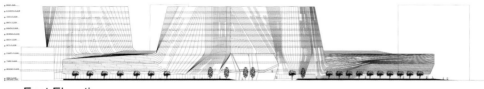

East Elevation

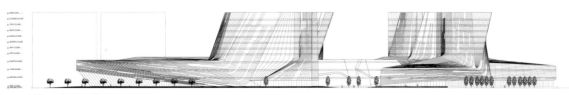

West Elevation

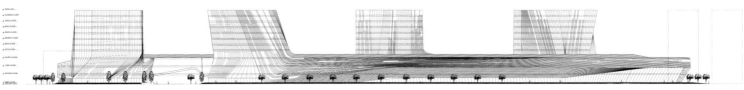

North Elevation

• New City Landmark — Mixed-use Architecture

Architects · JDS Architects

Hangzhou, China
Hangzhou Xintiandi Commercial Center

杭州新天地商务中心

Hangzhou Xintiandi Commercial Center locates on the site on East Street in north Lower City with its east extending to Hang Xuan Railway, south to Chang Dawu Road that is in project, and west to new North Shixiang. 6 km from Wulin Square and 7.50 km from The West Lake, it enjoys a convenient transportation; its west comes to east New River, Shang Tang, north to the Half Mountain, getting access to a extraordinary natural views.

To meet the demand arising from urban functions and social development, Hangzhou Government proposed a relocation of heavy machine factories. Given a name as Hangzhou Xintiandi Commercial Center, the site is planned to be established as a area that features a internationalism city business center and cultural innovation, multi-function gathering together recreation, culture, and business and leisure headquarters. Finally it will be the largest city complex area in Hangzhou.

On the site of north Xintiandi and boasting self as a highly recognizable building on Shixiang Road and east new East Road, this project is a symbolic building on the outside of the whole Xintiandi project. As part of Xintiandi project, meanwhile it is not so weird to catch people's eyes relying on its strangeness. Taking the high requirement to rooms' environment in a 5-star hotel into consideration, designers start from its functions and throw all the useless ornaments, so every part of this building fully displays the best function characteristic. While they design the modern building the designers incorporate

New City Landmark — Mixed-use Architecture

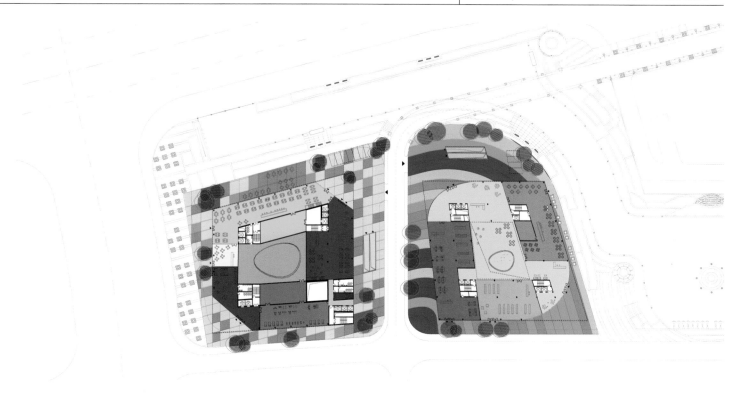

新天地区域鸟瞰图

the living concept of traditional siheyuan. That's the reason why such a tranquil and comfortable atrium context appeared. Also considering the practical demand to both investors' and hotel's operation, structure design in the project pursued a simple, stable, secure, and fast construction with a reasonable cost.

Building's plane shows us a siheyuan structure. Its west part is lower and gradually rises along its way to east part. Atrium sits on the third floor on hotel, and below it there is a double height swim pool. Lobby located at the east of ground floor, connecting restaurants and a gym center. Hotel offices and the rest of the restaurants are on the second floor. The third floor provides a banquet room and conference room. The forth

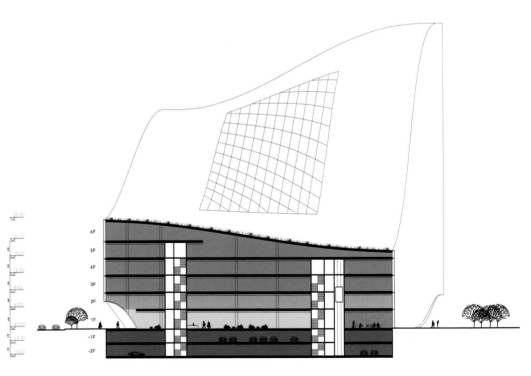

floor and above are all guest rooms and the top floor is the site for various facilities.

Northeast corner of this area has an entrance to cars and a turnaround, plus some car spots. Entrance of underground parking is on the west side. To provide more ground area and public space, two corners of the building are upwards, allowing autos and pedestrians coming under them. The west of the site locates a outdoor's café, connecting the restaurant that is on the north side of the ground floor, which offers visitors a outdoor opening space to enjoy river landscape.

　　杭州新天地商务中心项目位于下城区北部东街道，项目东至杭宣铁路、南接规划中的长大屋路、西靠东新北石祥。距武林广场约6km，距西湖约7.50km，交通便利。西临东新河、上塘，北朝半山，周边自然环境优越。

　　为顺应城市功能和社会发展的需要，杭州政府提出将重型机械厂搬迁以为顺应城市功能和社会发展的需要，以"杭州新天地商务中心"为名，把本地块建设成一个集文化娱乐、商业休闲等多功能于一体的国际化城市次级商贸商业中心及文化创意园区，建成后将成为杭州最大的城市综合体。

　　本案位于新天地北区，在石祥路及东新东路上辨识度极高，是新天地整体项目位于外部的标志性建筑，同时作为新天地项目的一部分，又不应过于怪异，哗众取宠。考虑到五星级酒店对客房环境的高要求，设计从功能出发，摈弃所有无用的装饰，使建筑的任何一部分都在功能上有着充分体现。在设计现代化建筑的同时，吸取传统四合院的居住理念，从而设计出安静怡人的中庭环境。考虑到投资方及酒店运作上的实际需求，本案的结构设计简单稳定，能够安全、迅速的建设并保证造价合理。

　　建筑平面上呈四合院结构，西部偏低，并逐渐向东部升高。中庭位于酒店3楼，中庭之

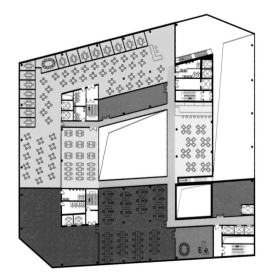
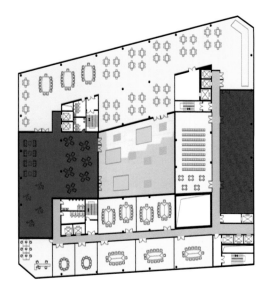

下为双层挑高的泳池。大堂位于一楼东侧，周围与餐厅、健身中心相连。酒店办公室及其余餐厅位于二层；三层为宴会厅及会议室；四层以上全部为客房；顶层为设备层。

地块东北角设有车行入口及回车场，并设有部分车位，地下车库出入口位于西侧。为营造出更多地面公共空间，建筑两角上翘，使得机动车和人流可从其下经过。地块北侧为户外咖啡座，可与一楼北侧餐厅连通，为欣赏河景的游客提供了一个开放的户外场所。

• New City Landmark — Mixed-use Architecture

New City Landmark — Mixed-use Architecture

Architect · HMA
Cooperate Architect · Huayang City, Suzhou Architectural Design Institute Co., Ltd.
Dayun City, China Building Technique Group Co., Ltd. (Suzhou)
Area · Dayun City: 260,000 m²
Huayang City: 115,000 m²

Suzhou, China

Suzhou Dayun City & Suzhou Huayang City

苏州大运城 & 苏州花样城

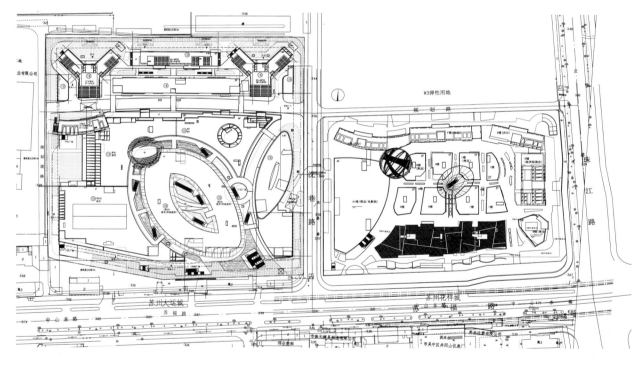

This project is the very first time-consumption shopping park with a 400, 000 m² area in Suzhou. It is the largest business synthetical block constructed in Mudu town by HMA, a gorgeous finishing for modern business culture of Mudu.

Commercial Environment

The east and west of Suzhou structure comprises main Suzhou downtown, industry developement, and new hi-tech development, with the Jianggan Road as its axis. Along with the rapid progress of the expansion construction of City Rings, a western great city commercial circle will emerge in the west area of Suzhou. In this commercial circle professional market on Changjian Road, Xujiang New City, Jinfeng innovation development, and Mudu historic town serve as important points.

Mudu town, one of the 12 districts of Guhua and business function areas and located in the center of this circle, will show a bright prospect in modernized real estate, modern service, and modern manufacture. Dayun city and Huayang city, with their sitting in great business circle of westen Suzhou, is destinated to lead business pattern's forming in western Suzhou, creating a totally new business level.

Project Position

The project will be planned to the commercial heart of western Suzhou, allowing a group for large and mutiple property form there, which can level up commercial culture in Wuzhong and add radiation and attraction of the sub-center of this city.

本案是苏州首例400 000 m²一次性消费型购物公园。HMA 在木渎镇打造苏州西部最大的商业综合体（花样城、大运城），为木渎的现代商业文化增添绚丽的一笔。

商业环境

苏州主城、工业园区、高新区共同构成以干将路为主轴的东西大苏州格局，随着城市环线拓展建设的快速进程，苏州城西版块以长江路专业市场、胥江新城、金枫创意园、木渎古镇商圈为重点的商业带将形成四位一体的城市西部大商圈。其中位于大商业区的中央片区，作为苏州市固化的12个区域及商业功能区之一的木渎古镇，未来发展方向将是以现代房产业、现代服务业、现代制造业为主。 地处城西大商圈CBD商业带，大运城、花样城必将统领苏州西部商业格局，创造城西全新的商业高度。

项目定位

本项目定位为苏州西部的商业核心，成就苏州西部大规模多物业形态集群，提升吴中地区商业文化，增加城市副中心的辐射力和吸引力。

Dayun City

Comprehensive Arrangement

This project consists of Haosheng hotel, Da Runfa supermarket, Zhenbing skating place, large clothing shops, and SOHO offices. And its commercial space encompasses two overlap malls. That is a really brilliant idea. There are four districts.

District A: building No.1, No.2 and No.3 are high-rise official building, with a height of 80 m and a floor height of 5.8 m.

District B: a main store of large fashion retails

District C: from the ground floor to the forth floor and fifth floor, some portions locate experience consumption center.

District D: from the ground floor to the forth floor, main restaurants sit there, and from forth floor to fourteenth locates Haosheng hotel.

Commercial design

Coming from "time comsumption experience", the project features the interaction and organization between moderate moving knote settings and primary and secondary moving lines, creating an interesting and amiable commercial atmospher. The manipulation of convinient and smooth liner design as well as comfortable and fashion spacial design bring forth a passionate business space. By plane arrangement, natural lights and water go through commeicial space as a key generatrix. Then shopping atmosphere design derives from visual and spatial comfortablity, emphasizing on consumers' subjective requirement and enhancing commercial environment quality from muti-perspevtive. All these guarantee the project as a new identification of this region.

Elevation design

The entire elevation design is a new interpretation, understanding and analysis of the "region at the south of Changjiang River" culture by modern architects. During the design process, the designers absorbed the brilliance of the structures' form and traits, folk culture, classical poems and lyrics, and local culture, integrating them into structure design, especially elevation expression. This is a wonderful interpretation and practice of Suzhou modern culture and consumption idea.

大运城：

总体布局

本项目由豪生酒店、大润发超市、真冰溜冰厂、大型服饰主力店、SOHO 办公组成，商业部分由两个 mall 巧思叠加而成。总共分为 A、B、C、D 区。

A 区：1~3 号楼为高度 80m 的高层办公区，层高 5.8m。

B 区：设置成为大型时尚品牌零售主力店。

C 区：一至四层、五层局部设置体验式消费中心。

D 区：一至四层为餐饮主力店，四至十四层为豪生酒店。

商业设计

本项目从"一次性消费体验"的理念出发。适度的动线节点设置以及主次动线的交替组织互动，充分营造出多趣、亲切的商业氛围。通过便捷、流畅的流线设计、舒适时尚的空间设计，创造富于激情的商业空间。从平面布置、将自然采光、自然流水贯穿商业主要动线。购物氛围从视觉及空间舒适度等方面入手，着重考虑顾客主观使用要求，多角度地提升商业环境质量，使之成为整个区域商圈的新标杆。

立面设计

整个建筑立面是现代建筑师对"江南文化"的全新诠释、理解和解构。设计过程中萃取江南地区的建筑形态、建筑色彩、曲艺文化、诗词歌赋、当地人文等诸多元素，将其融入建筑设计尤其是立面表现之中。这是对现代苏州文化和现代消费理念的诠释和实践。

Huayang City

Comprehensive Arrangement

Huayang City is made up of A — I nine buildings, striving to create a rich, interesting and comfortable shopping and tour experience on the commercial blocks. There are three parts: 1. Xingmei cinema and shopping center on side of Chenxiang Road. 2. four office buildings on both Zhujiang Road and north side of a road that is in planning. 3. four pedestrain malls with "modern regions having the south of Changjiang" features. Introduction of them are as follows:

1. shopping and relaxation center

located on one side of Shenxiang Road, building A is the shopping center including retail, commercial, relaxation and leisure facilities. Its body has three floors: the ground floor and the second floor provide retails, and the third floor is Xingmei cinema; underground floor is retails which connects the underground floor of Dayun City by a underground pedestrain mall.

2. office facilities

The east and north of base are used as SOHO offices that are the most popular in market, with the height of 100m. The forth floor and the below podium serve as a really fine pedestrain mall.

3. characteristic commercial and fashion pedestrain mall

The central part is designed as a pedestrain mall, in which fine goods shops, large and middle-scale restaurants, and theme experience stores are arranged there.

花样城：

总体布局

"花样城"由A~I共九栋建筑组成，力图营造街区商业的丰富趣味性和舒适的购物游玩体验。功能分为三个部分：沿沈巷路一侧设置星美影院及购物中心，沿珠江路和北侧规划道路布置四栋办公楼，以及地块中部布置四栋具有"现代江南"特色商业步行街。

① 购物休闲中心

沿沈巷路一侧的A栋为购物中心，以零售精品商业休闲娱乐设施，主体为三层。一、二层为零售精品商业，三层为星美影城。地下一层为零售精品商业，并可通过地下步行街同西侧的"大运城"的地下一层相连互动。

② 办公设施

基地北侧及东侧为设置目前市场热销的SOHO办公，总建筑高度为100m。四层以下的裙房部分设置为精品步行街。

③ 特色商业和时尚步行街

中心板块为步行街设计，业态布置为特色精品商铺、大中型餐饮主力店、主题体验店。

The difficulty in this project is how to endow these two synthetical buildings of giant volume that are so close with their own particularities whilst they share a harmonious enviroment. Thanks to the diversified architecture, volume, business running styles, commercial mode arrangement, there is no race between them, and what's more, at the same time, they complement each other resulting in their both progress.

The commercial pattern of Huayang City is "Town" style, which combines exclusive Suzhou culture with mini streets feelings in the region on the south of Changjiang River, creating a town mall center brimed with humanity. The commercial pattern of Da Yuncheng is "double Mall" style. It enjoys natural water system and sunlight element suffusing the whole shopping context. And the indoor mall locates several indoor

　　本项目的难点在于如何使二个如此近距离的大体量商业综合体既具自身特色，但又能很好地融合共处。花样城、大运城以不同的建筑风格和建筑体量、不同的商业运营模式、不同的商业业态排布设置，避免了二者的相互竞争，却又相互弥补不足，使二者齐头并进。

　　花样城为"Town"型商业，设计上既结合苏州特有文化，又加以江南城市小街的风情，

experience centers. Though seperated by water system naturally, the two malls interact each other by a rich and various moving lines connections.

The large main stores in the two project scatter with a certain order, smartly avoiding their confrontation, and additionally, they complement each other. This idea allows thses two project can be seen as a integrate body or individuals, generating an all-rounded commercial patterns, complementary commercial environment, and harmonious effect.

打造成一个亲切的小镇式的购物中心。大运城为"双Mall"型商业，室外mall拥有水系、阳光的自然元素贯穿整个购物环境，室内mall拥有多项室内体验中心，二者虽以水系作为自然分隔，又以丰富多变的动线连接互动。

两个项目中的大型主力店的有序而又错落设置，避免业态的相互冲突，做到相互弥补共生。使得两个项目可分可合，业态面面俱到、商业氛围相互依托、两个地块达到和谐共生的效果。

- New City Landmark — Mixed-use Architecture

New City Landmark — Mixed-use Architecture

Architect · Llewelyn Davies

Xiangyang, China

Cultural Complex

文化综合体

The project is a cultural complex, located in the new district of Dongjin in Xiangyang, Hubei province. As an important cultural building of the new district, the city library has been designed as a cultural center to meet the needs of the development of city. In guaranteeing the main functions of "collecting, borrowing, reading, consulting, relaxing", designers integrate leisure, culture, performance, entertainment and communication as a whole, and create the traditional library into a cultural complex.

The cultural complex consists of the following parts: library area, recreational facilities, theatre, book mall(books trading center), art and culture exhibition hall, education center, office supporting and underground garage.

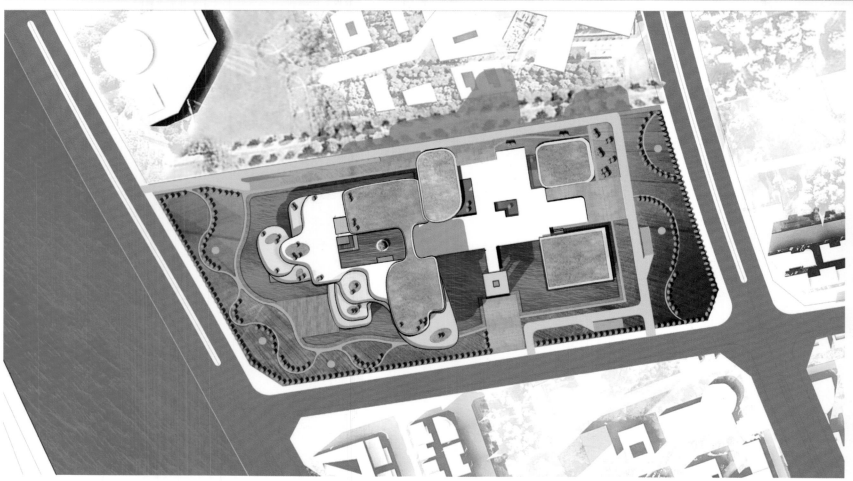

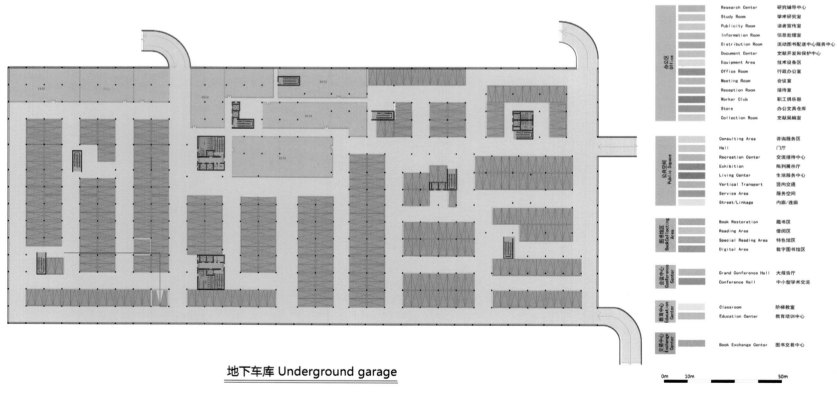

地下车库 Underground garage

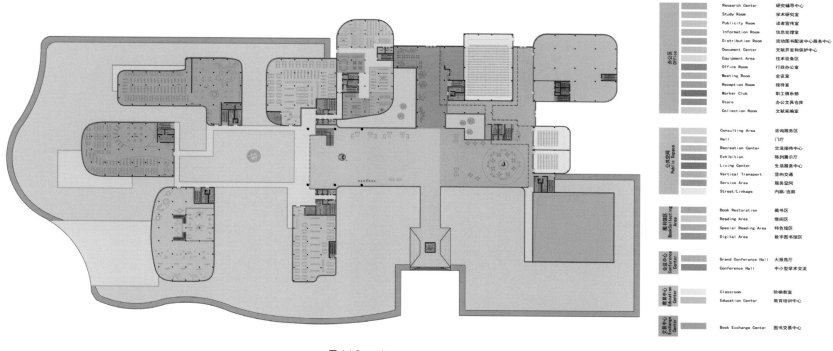

一层 L1 floor plan

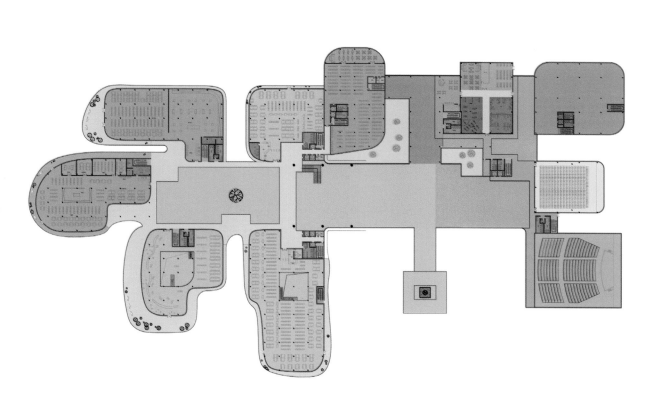

二层 L2 floor plan

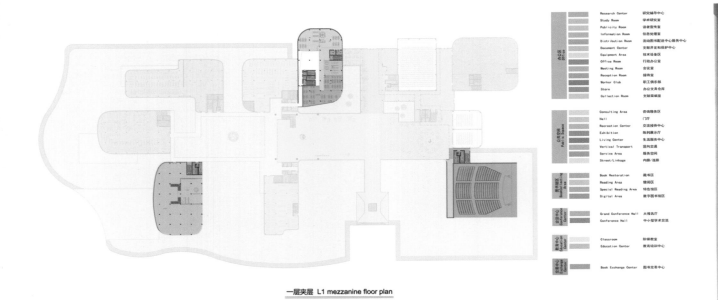

一层夹层 L1 mezzanine floor plan

该项目为文化综合体。项目位于湖北省襄阳市东津新区。作为重要的新区文化建筑，满足城市发展的需要，城市图书馆被作为文化中心进行设计，在保证图书馆"藏、借、阅、咨、休"主要功能前提下，更是将休闲、文化、表演、娱乐以及交流等综合功能整体化设计，将传统意义的图书馆打造成为文化综合体。

该文化综合体由以下几部分组成：图书馆区，休闲设施配套，剧院，书城（图书交易中心），艺术文化展馆，教育中心，办公配套和地下车库。

- New City Landmark — Mixed-use Architecture

Architect · LWK & Partners
Area · 3,160,000 m²

Guangzhou, China

Asian Games City

广州亚运城商业体

Asian Games City is broadly divided into three subject areas: semi-outdoor pedestrian area, riverside retreated dining area and department store area which will be rebuilt from the original media center. The various levels are connected by bridges and platforms so that the circulation will not be obstructed.

The commercial streamline design cooperates fully with the various transport facilities: the planned subway from the basement directly connects the main southwest entrance of the mall, the bus terminus next to the south door of the Asian Green Square; small pier at the eastern riverside, can be used for the tour boat mooring; the northern commercial street can be used for the traffic from international zone and the northern residential area; coupled with northwest and northeast huge entrance plazas, absorbing the flow of people both from the land and river.

The commercial design features the communion of the multivariate space while the pedestrian street provides round-the-clock shopping space with natural ventilation and sunlight. One end of the pedestrian street is located in the center of the commercial outdoor plaza with the best riverside landscape. Along the riverside is the outdoor deck dining area. All kinds of platforms connecting various heritage and natural attractions of the Asian Games, brings the lively atmosphere of outdoor area into the traditional mall space. Both the dining area and the ground floor are blended with the riverside green belt and the use of water features brings the natural waterscape design into the dining area.

The gateway is a landmark of the Asian Games. The gateway design is not only architectural highlight, but also provides 3,500m² of banquet clubhouse, swimming pool and other facilities for the commercial body. The Asian Heritage Museum located in the vertical transportation core of the gateway with vertical greening will become a tourists attraction, continues the Green spirit of the Asian Games. At the bottom of the gateway, the footbridges planted with trees are the most important green belt of the entire project. For commercial considerations, it was designed to keep the main flow line at the mall interior, while the Bridge of the Asian Games has become the role of Green Park and the buffer between commercial and southern residential area, giving a new meaning for the heritage.

New City Landmark — Mixed-use Architecture

商业体大致分为三个主题区域：半户外步行街区、滨江退台式餐饮区与改建自原媒体中心的综合百货区。各部分于每一层均以天桥和活动平台连接贯通，令四方人流动线畅通无阻。

商业体的流线设计充分配合各种交通设施：规划中的地铁四线从地下一层直接连通商场西南主入口，公交车总站设于南面亚运之门绿化广场侧；东面江边设有小码头，可作游览船停泊之用；北面商业街车道吸纳于国际区与北面住宅的车流；加上西北和东北两角均设有巨大的入口广场，尽揽水、陆两路人流。

商业设计的特色是内外共融的多元空间，步行街提供自然通风与光线的全天候的购物空间，步行街的一端是位于商业体中央的户外活动广场，广场饱览基地最佳的滨江景观，视线往

江边移是退台式餐饮区，各式活动平台连接各个亚运遗产和自然景点，把室外活动气氛和区域的特色带进传统的商场空间。餐饮区与地面层与滨江绿化带融合，利用水景设计巧妙地配合滨江意境，天然的水景设计就像把江水带进餐饮区。

商业区的门户是具地标性的亚运之门，门框设计除了成为建筑亮点外，也为商业体提供了 3 500 m² 的宴会会所，游泳池等设施。会所下部把出口的垂直交通核心与亚运文化博物馆融合，以垂直绿化为设计重点，此公共性令亚运之门成为游客必到的景点。延续亚运的绿色精神，亚运之门底部，原有连接各个亚运遗产的行人天桥被打造成亚运森林，是整个项目最主要的绿化带，一直延伸到江边。为了商业考虑，设计保持主要的人流动线于商场内部，亚运之桥成为了绿化公园的角色，成为商业与南住宅区的缓冲区，为遗产赋予了新的意义。

- New City Landmark — Mixed-use Architecture

Wenzhou, China

Yongjia World Trade Centre

永嘉世贸中心

Architect · UNStudio
Client · Shanghai World Trade (Shanghai) Holdings Group
Program · Retail, Office, Hotel, Residential, Exhibition, Cultural

The Yongjia World Trade Centre will create a new image for the WTC brand and will become a unique symbol for the new riverside city of Wenzhou. The World Trade Center is located in the Oubei Sanjiang Area which has the highest development potential in Wenzhou. According to the master plan, the Oubei Sanjiang Area in Yongjia is positioned as an integrated functional area with, among others, business, modern residence, tourism service, leisure and entertainment functions rolled into one. The functional shift of the WTC area, from a business and financial district to a mixed-use development which includes cultural and recreational facilities and a high percentage of residential properties, will create a forward-looking and sustainable city district that has all the components needed to support economic growth whilst propelling social

New City Landmark — Mixed-use Architecture

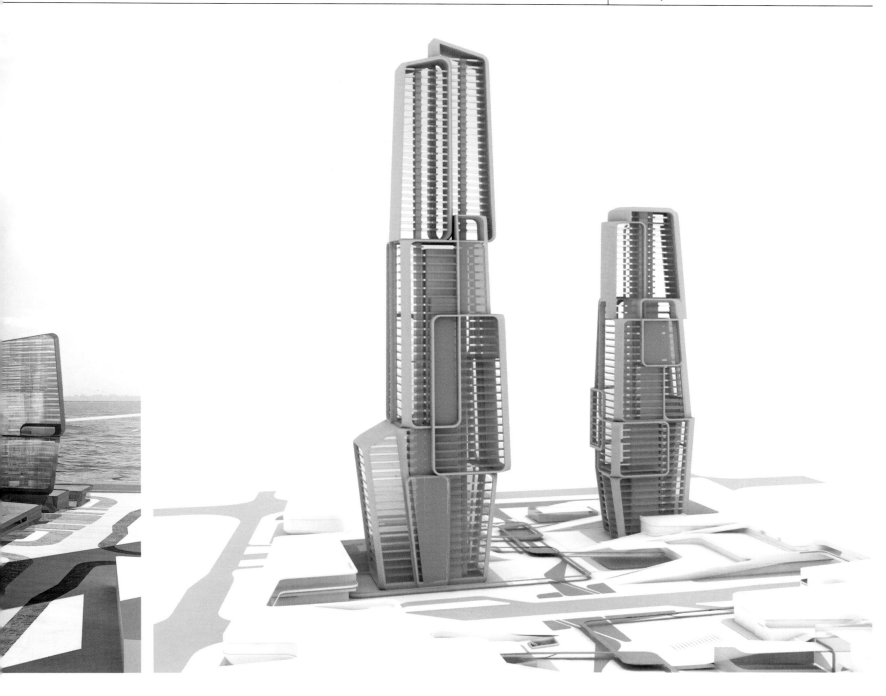

connectivity and local identity.

UNStudio's competition design proposed 5 towers ranging from 287m for the main tower to 146m for the smallest tower. The office towers, including the World Trade Centre offices, are situated in the North area of the development. High end residential apartments are located on the upper levels of these towers, with residences that enjoy the best 360 degree views overlooking the whole peninsula.

The programme mix of office and residential ensures an intertwining of functions and activation throughout the day and night.

In the South part of the development residential towers and a hotel tower are located. In accordance with the competition brief the total above ground area adds up to 500,000m², including shopping and commercial areas of 150,000m², office areas of 160,000m², hotel area of 50,000m² and a high-rise condominium area of 140,000m².

UNStudio's design for the new World Trade Centre presents a green neighbourhood in the sky which is combined with three main elements: trade and business, diverse programme mix and accessible public landscape.

The landscape is the unifying element in the overall design, providing the display element for the tower objects. It is in large parts publically accessible and establishes a continuous green connection that links through the central green axis to the riverfront area.

永嘉世贸中心将为WTC品牌带来一个新的形象,同时也将成为温州江边新城的独特地标。项目位于瓯北城市新区三江街道,是温州极具发展潜力的地区。根据总体规划,永嘉瓯北城市新区三江区域将被打造成综合商业区,集商业、现代化住宅、旅游服务及奢华娱乐功能于一体。世贸中心的功能转换,即从商业和金融区转变为结合文化休闲及高密度住宅的城市综合发展区,将创建一个具有前瞻性和可持续发展的城区,具备所有必需设施以支持经济增长,同时推动社会联系和本地特色。

UNStudio参与竞标的设计方案提议了5栋塔楼,最高的塔楼为287m,最矮的塔楼为146m。办公大楼内设世贸中心办公室,建筑位于选址的北面。这些塔楼高层被用作高端住宅公寓,住户能够360度俯瞰半岛全景。

办公和住宅的混合,交织的功能确保了建筑白天黑夜都有人员活动。

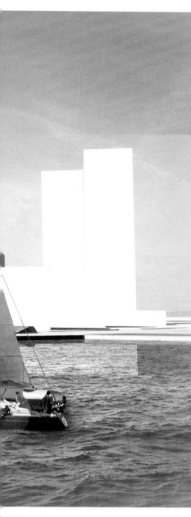

The competition design proposes zoning by differently themed areas embedded into the landscape that are related to cultural and entertainment functions. At the central water stream and on the lower podium level the landscape has a lively character with various functions. In contrast, the top of the roof enjoys a more private character. This zoning creates diverse green areas that can satisfy the different needs of residents, visitors and business people alike.

选址南面是住宅楼和酒店。与竞标方案相一致，总的地表建筑面积为500 000m²，其中购物中间及商业区域面积为150 000m²，办公区域面积为160 000m²，酒店面积为50 000m²，高层公寓面积为140 000m²。

UNStudio的设计呈现了一个空中绿色社区，结合了贸易和商业、多元化项目组合以及无障碍公共景观这三个主要方面。

Plan Level 1

TOWERS | CONCEPT AND IDENTITY

TOWER COMPOSITION

FRAMING - IDENTITY

DIFFERENTIATED NEIGHBOURHOODS

FACADE VARIATIONS

TOWER 'EYES'

　　景观是总体设计的统一元素，为每栋塔楼提供展示元素。大部分对公众开放，通过中央的绿轴将连续的绿地与河畔区域联系起来。

　　竞标方案提议了根据不同主题区域进行的分区，嵌入与文化和娱乐功能相关的整体景观中。中央的水景和低处的平台，各种功能性景观拥有生动活泼的特征。相比之下，建筑顶层的屋顶景观则拥有更加私人的特征。这种分区创造出各种各样的绿地区域，满足住户、游客及商业人士等的不同需求。

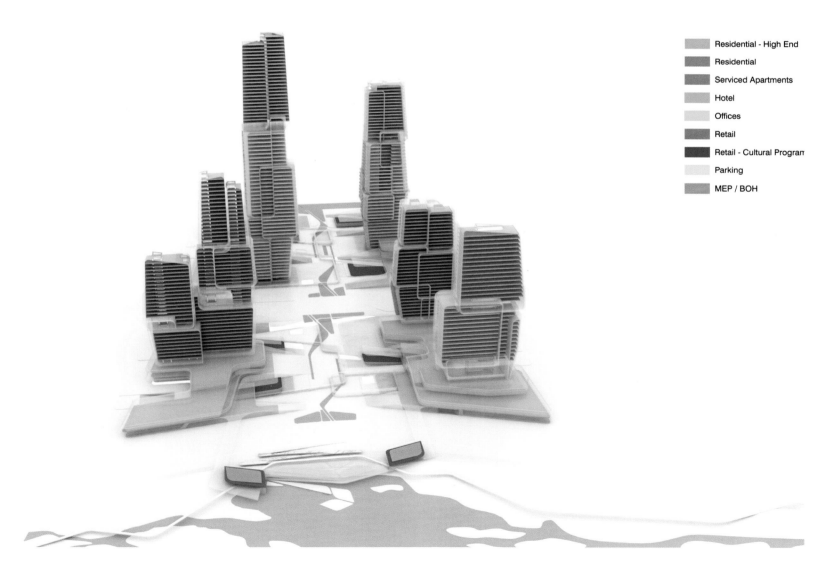

- Residential - High End
- Residential
- Serviced Apartments
- Hotel
- Offices
- Retail
- Retail - Cultural Program
- Parking
- MEP / BOH

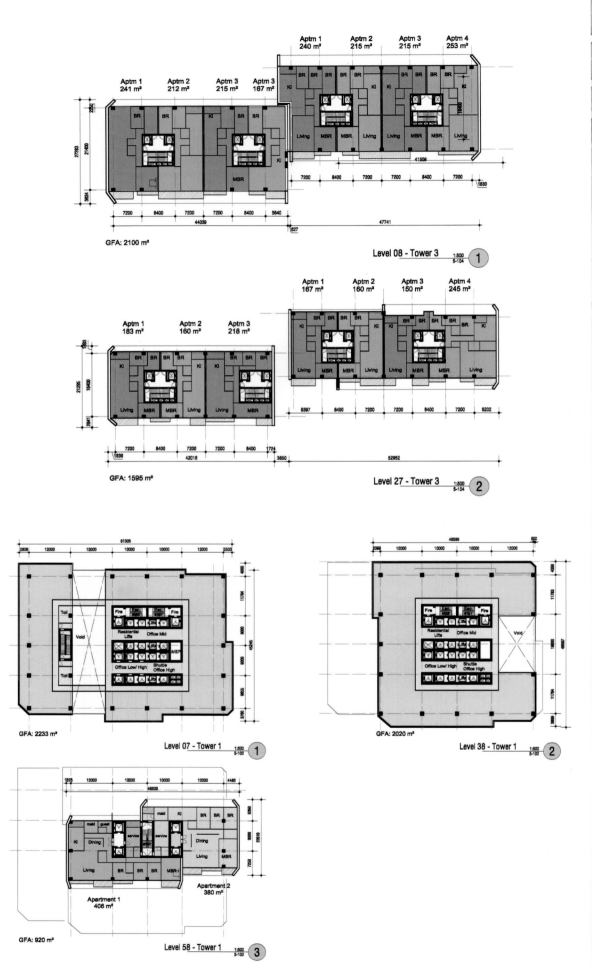

• New City Landmark — Mixed-use Architecture

Beijing, China

China World Trade Center

中国国际贸易中心

Architect · 5+design
Client · China World Trade Center Ltd.
Lead Designer · Arthur Benedetti
Area · 332,500 m²
Program · Retail, Entertainment, Residential, Office, Exhibition

China World Trade Center has always been associated with superlatives. It's where the country's first Starbucks opened, for example, and home to Beijing's tallest office tower. It's hailed as the capital's most prestigious mixed-use development when it was launched in 1990 with premium high-rise offices, luxury hotels and, inside China World Shopping Arcade, haute couture stores such as Fendi, Prada, Hermès, Gucci and Bulgari. Nearly a quarter-century later, 5+design has designed a multiphase expansion that will enlarge and improve the arcade, now known as China World Mall, and re-establish the trade center as the city's premier shopping destination.

The expansion will provide new luxury and mid-market retail and entertainment to complement and knit together the existing buildings and uses, allowing China World Trade Center to provide the latest in lifestyle, home, dining and entertainment choices. These venues will be housed in contemporary structures featuring streamlined facades of shimmering glass with metal and stone accents, and rooftop gardens with restaurants and bars offering views out to the city and back into the trade center campus. At night, translucent and transparent glass panels will be illuminated, turning the commercial landmark into a beacon along Jianguo Road and the 3rd Ring Road.

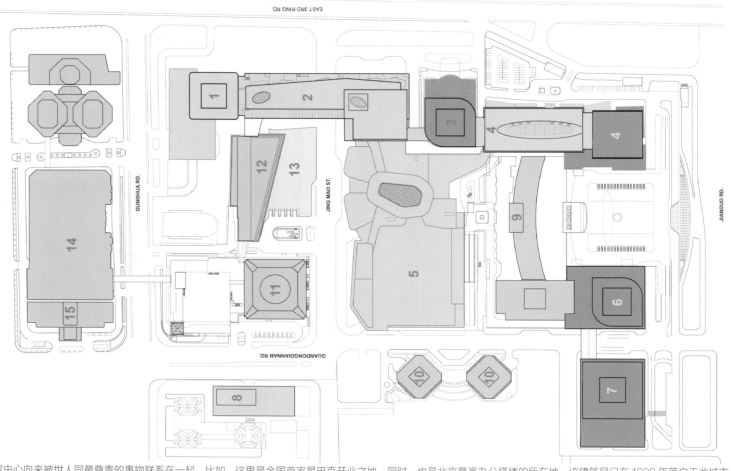

中国国贸中心向来被世人同最尊贵的事物联系在一起。比如，这里是全国首家星巴克开业之地，同时，也是北京最高办公塔楼的所在地。该建筑早已在1990年落户于此城市，拥有高质量的高层办公楼、豪华酒店，以及世界各地知名品牌购物商场和高级女装店，如芬迪、普拉达、爱马仕、古驰、宝格丽等，所以被誉为是首都最负盛名的综合体开发项目。将近25年后的今天，5+Design为此项目设计全方位扩展，使之成为更大型、更完善的商场，现在被称为是中国国贸商城，并新建贸易中心成为全市最大型购物目的地。

此次项目扩建将提供奢华而又适合中端市场的商业及娱乐体验，与现有建筑相互补，使中国国际贸易中心为人们提供最新生活、家居、餐饮、娱乐等不同选择。这是一座极具现代化的建筑，外立面拥有金属闪石玻璃的流线型。屋顶花园设有餐厅与吧台，为人们提供对这个城市不同角度的景色观赏。到了晚上，半透明和全透明玻璃板亮起，使该商业地标建筑瞬间转变成一座为建国路与三环路照明的灯塔。

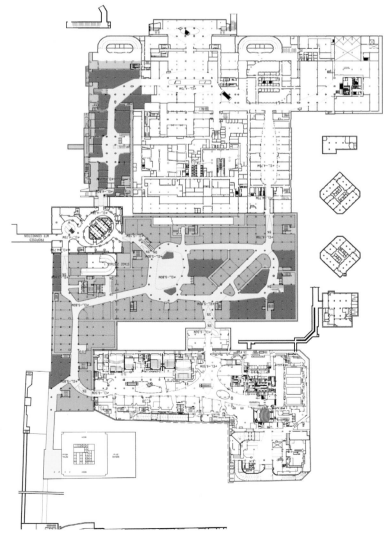
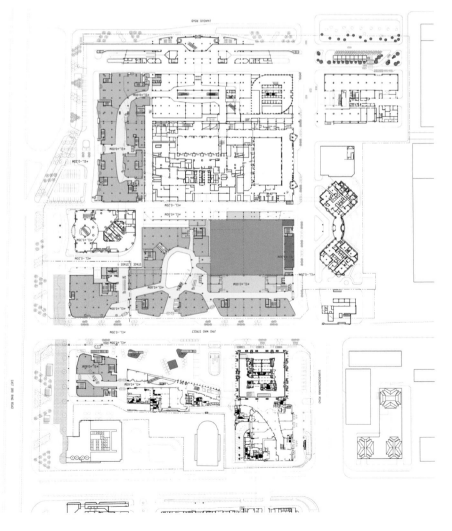

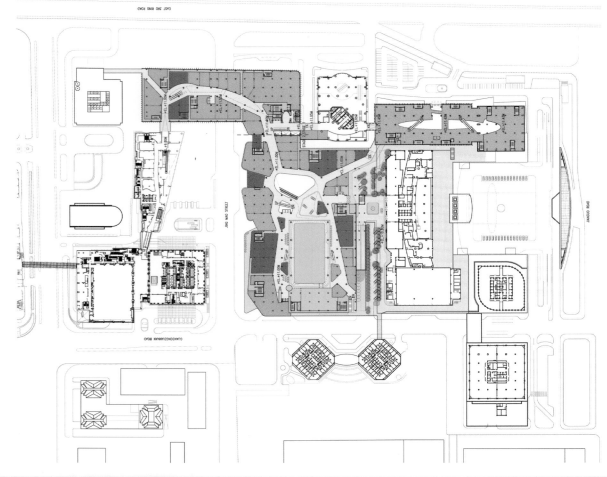

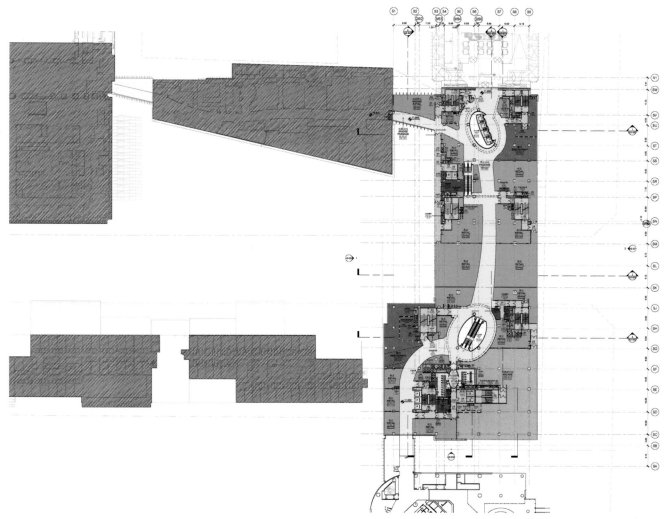

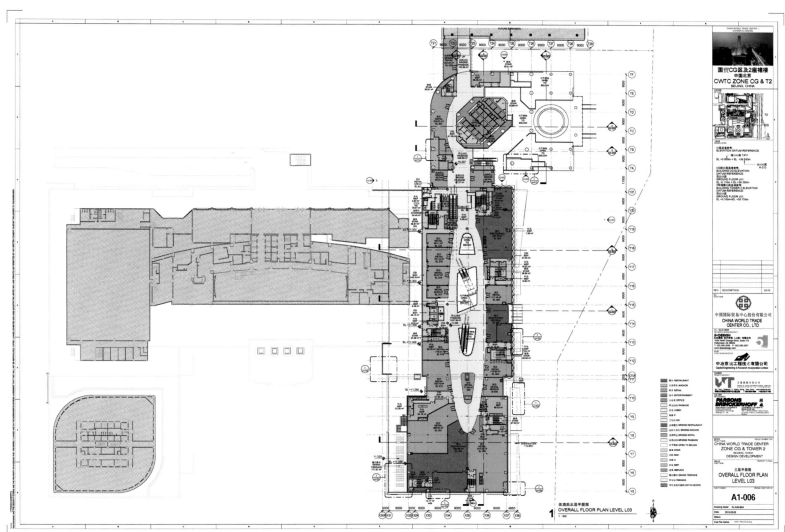

- New City Landmark — Mixed-use Architecture

New City Landmark — Mixed-use Architecture

Dongguan, China

Dongguan International Business Center

东莞民盈国贸中心

Architect · 5+design
Design Team · Tim Magill, Chris White, Ramón Hone (designer towers)
Client · Dongguan Minying Plaza Development Co. Ltd.
Area · 1,049,237 m²
Program · Retail, Restaurant, Hotel, Office, Entertainment, Exhibition

Dongguan, one of China's leading manufacturing centers, has embarked on an effort to transform its image from factory town to modern consumer oasis. One of the most significant steps in the makeover is Minying Plaza, a master-planned mixed-use development in the heart of the city. Five towers – one hotel and four office and/or residential buildings – will rise above a five-level "leisure valley" of international retail stores and restaurants. The project will enhance community life in other ways with the inclusion of a health club, ice skating rink, children's playground, 15-screen cinema, rooftop amphitheater, adult education center and local history museum.

Like Dongguan's lush boulevards, Minying Plaza's architecture is intended to evoke the beauty and serenity of nature. Walls, terraces and plazas made of different kinds of stone will be softened with lush vegetation, fountains and waterfalls. The Cloud, a white structure "floating" overhead, will provide shelter from sun and rain. Glass bridges spanning the retail corridor will afford safe, unobstructed views. And at a height of 388 m, the tallest tower will soar as a metaphor for Dongguan's city flower, the Michaela, its curvilinear glass walls wrapping the spire like petals around a delicate bud.

	Public Plaza and Space 公共广场和空间
	Green Roof Terrace 屋顶花园
	Experiential Open Space 体验性开放空间

东莞长期以来一直被视作中国最大的外国商品生产地，而今它希望把自己建设成为一个展示国人自主构思设计产品的橱窗。该转变的重要一环是民盈广场，即位于城市中心的一个总体规划的综合体开发项目。五栋大楼为一栋酒店和四栋办公或住宅大楼，将矗立在一个由国际零售店和餐厅组成的五层"欢乐谷"之上。项目将以其他便利设施提高社区生活，包括健康俱乐部、溜冰场、儿童游乐场、15屏的电影院、屋顶的圆形剧院、成人教育中心和主题为东莞商业史的展览馆。

类似东莞茂密的林荫大道，广场建筑试图唤起自然的美丽和宁静感。郁郁葱葱的植被、喷泉和瀑布给由不同种类石头打造的墙壁、露台和广场增添了柔和感。空中一个白色结构的"浮块"，有如白云般，用于遮挡阳光和雨水。横跨零售走廊的玻璃桥将提供安全、视线通畅的景致。最高的塔楼高达388m，有如东莞的市花翱翔在空中。米凯拉的曲线玻璃幕墙包裹着尖顶，宛如花瓣围绕着一个微妙的蓓蕾。

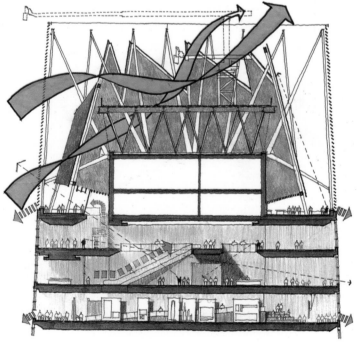

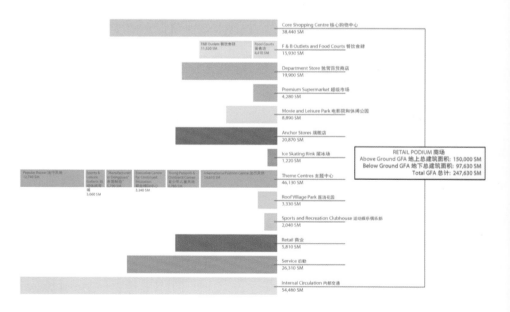

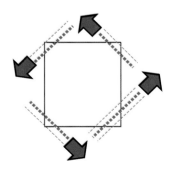
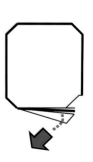
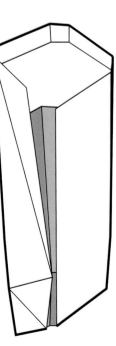

- New City Landmark — Mixed-use Architecture

Architect · 10 DESIGN
Design Team · Ted Givens, Adam Wang, Adrian Yau, Audrey Ma, Peby Pratama, Yao Ma
Client · Galaxy Group
Area · 304,285 m²
Program · Hotel, Serviced Apartments, Conference Centre, Commercial

Changzhou, China

Galaxy Moonbay Renaissance Hotel and Mixed Use Development

星河澜月湾万丽酒店及综合发展项目

The project is located in the West Tai Lake's Ecological and Leisure area, positioned to be one of the most important vacation resorts in the Delta Region of Yangtze River. The 310,000m² of development will form a modern leisure and commercial community, providing an iconic luxury hotel, serviced apartments, conference centers, offices and public spaces. Construction has started at the end of 2012.

"Given the unique geography of the site, situating along the 14 km coastline of Lanyue Bay, the client has given designers a vision to create a garden setting environment in order to help increasing the public spaces surrounding the lake", says Ted Givens, Design Partner of 10 DESIGN (often referred to as "10").

The masterplan design was inspired by the local craft of bamboo carving and the aerodynamic shapes of modern boat design. The buildings are all curved to reflect the fluidity of the lake.

Architectural design and planning arrangement respond both to its surrounding environment and the site itself. Careful consideration was paid to zoning and building arrangement within the site to maximize views and ventilation to all buildings, while at the same time retaining the existing view & breeze corridors. The concept of green design is further implemented via the introduction of green spaces above ground plane.

The MLP design is centered around 6 main design objectives:

1. To create an iconic high rise building facing the lake.

2. Arranging the buildings so that they do not block the view of the lake from properties located behind/north of the site.

3. Minimize impact of shadows cast by towers on sites to the north of the MLP site.

4. To seamlessly integrate the retail and F+B facilities with the existing park to the east to help support the existing park and pull visitors into the MLP site.

5. To provide each of the new residential units with views of the lake.

6. To maximize the possibility of cross ventilation for the buildings and to maintain lake breezes in all the garden areas.

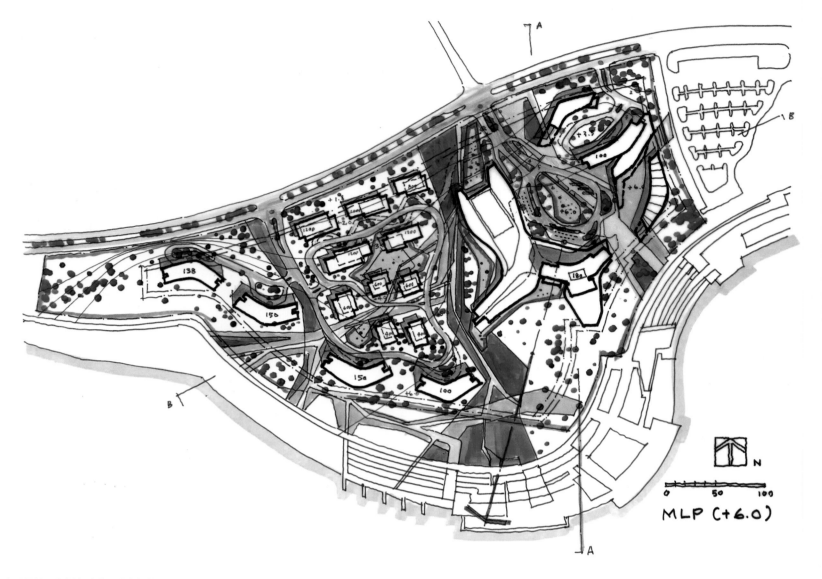

本项目位于常州市武进区滨湖新城的西太湖生态休闲区，将成为长江三角洲一个重要的旅游度假区。项目总面积达 310 000 ㎡，地块规划将形成一个现代化的休闲及商业社区，发展业态包括一间标志性的豪华酒店、酒店式公寓、会议中心、办公楼及公共空间等。项目已在 2012 年底动工。

其文特-10（拾稼设计）设计合伙人说："由于项目地处独特的地理位置——沿揽月湾 14km 长的海岸线，客户与我们分享了他们对本项目的愿景，激发了我们以花园环境提升滨湖公共空间品质，来打造本项目的灵感。"

此总体规划设计的灵感来自当地竹雕刻工艺及具有流线外形的现代船舶设计。建筑物呈

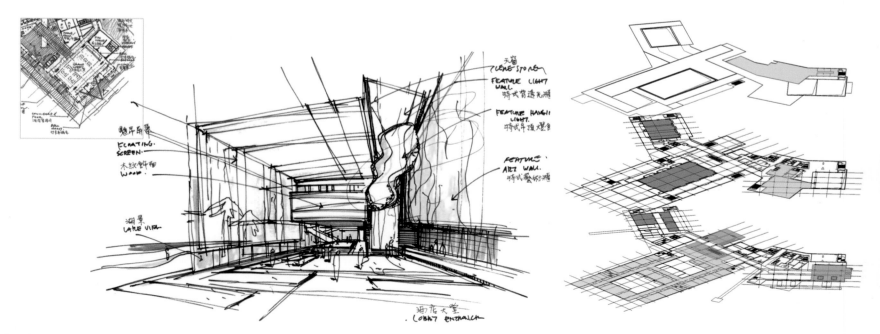

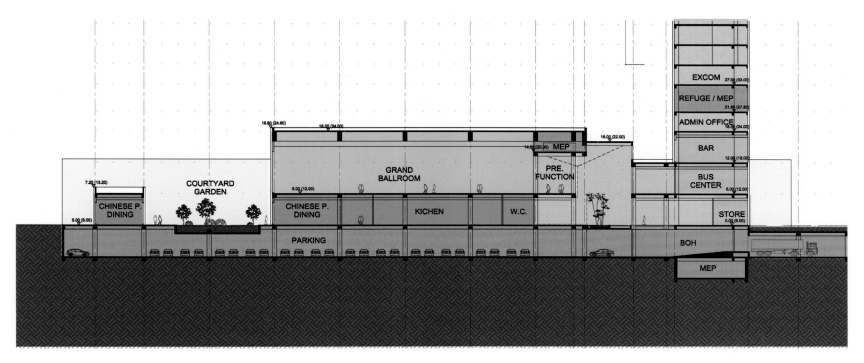

弧形设计,以反映湖泊的流动性。

建筑设计及规划布局与周边环境及基地本身互相呼应,相得益彰。设计细心考虑了基地分区和基地内建筑物的布局,以获得最佳的景观效果及自然通风,同时亦保留现有景观及通风迴廊。通过进一步引入绿地空间彰显绿色设计的概念。

总平面的设计围绕6个主要的设计目标:
① 创建一个临湖的标志性的高层建筑。
② 合理安排楼宇位置,避免其相互遮挡,以便使处于基地北部的楼宇均能享受到良好的湖景。
③ 将建筑物的阴影遮挡减少到最小。
④ 将基地内的餐饮购物设施与东侧的公园实现"无缝对接",既可以补充完善公共空间的服务设施,亦可将游客吸引至基地内。
⑤ 为每一个新建的住宅单位提供优美的湖景。
⑥ 最大限度地保持建筑物的自然通风,使湖面的微风吹进花园的各个角落。

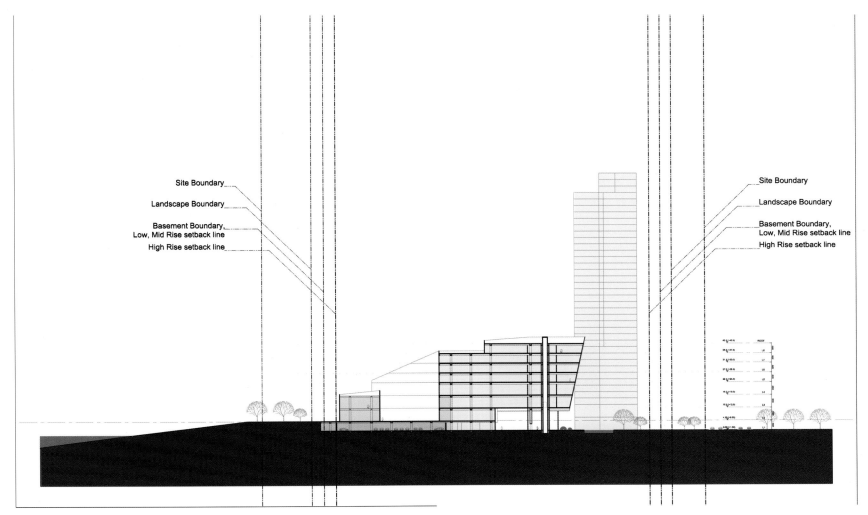

sales pavilion
售楼中心

headquarter office
总部办公

conference and business centre
会议及商务中心

hotel
酒店

• New City Landmark — Mixed-use Architecture

New City Landmark — Mixed-use Architecture

Architect · ARCODEC-COM Architectural Studio
Client · A local Commercial Agent, Owner of an existing open air Agricultural Market
Area · 58,000 m²
Program · Commercial, Parking, Health, Entertainment, Offices, Residential

Moldova, Romania

Aem Compl

Aem Compl

In 2012, the Client, who is the owner of an existing Open Air Agricultural Market in Chisinau, intended to build a residential district instead of the existing market, the Architect (Ion Eremciuc) suggested to make a mixed use complex which will include a commercial zone, a parking zone, a recreational zone and a living zone. Also he proposed to use the commercial building zones roof as green, recreational and active sport use space.

In order not to brake the access of the sun to the neighborhood building the architect designed 2 separate apartment buildings, 18 floors and 24 floors.

2012年，客户是基希讷乌一个现存露天农产品市场的业主，他想建一个住宅区取代现存的市场，建筑师建议客户建造一个混合功能的综合体，包括商业区、停车区、休闲区和生活区。

他还建议使用商业建筑区的屋顶作为绿色休闲和运动空间。

为了不阻碍太阳光线进入邻近建筑，建筑师设计了两个独栋公寓，分别是18层和24层。

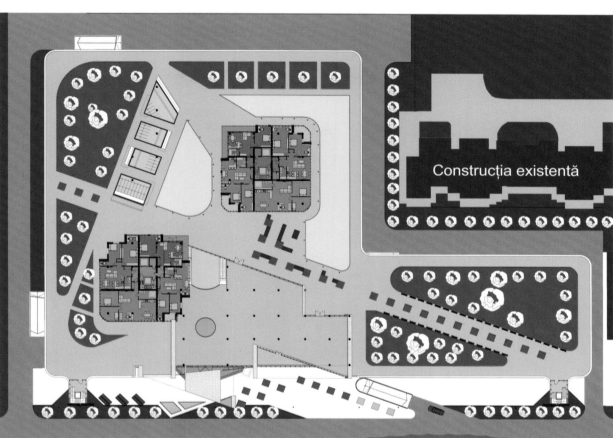

Construcția existentă

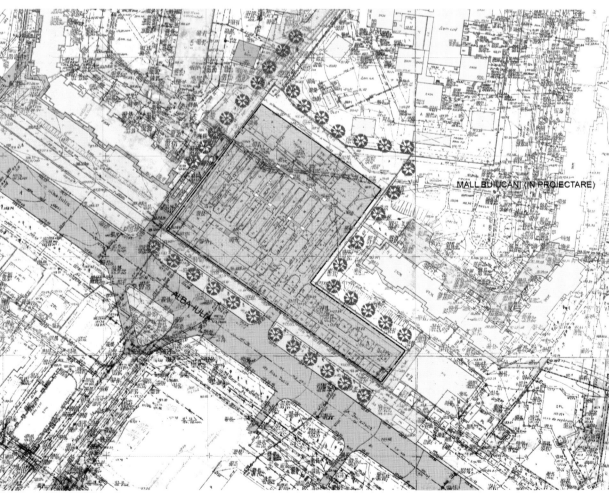

MALL BUIUCANI (ÎN PROIECTARE)

ALBA IULIA

• New City Landmark — Mixed-use Architecture

Architect · ATENASTUDIO + Archmaster studio
Client · Wuxi Taihu Pearl Estates Development Co., LTD, Mr. Cui Yang; Cheng Hsiung International Corporation, Mr. Chen Gang
Area · 561,411 m²
Program · Retail, Hotel, Entertainment, Cultural, Residential, Parking

Wuxi, China

Wuxi Masterplan: Mixed Use Building Complex

无锡综合体总体规划

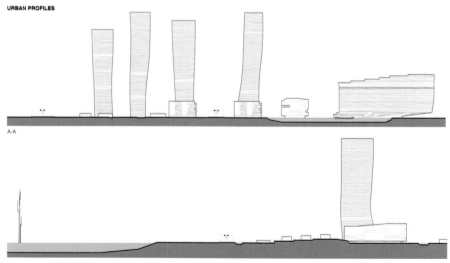

ATENASTUDIO in collaboration with Archmaster studio has developed a masterplan for a new district in Wuxi, China.

The masterplan project called "CONCEPT DESIGN FOR BUILDING'S PROJECT – LOT A.02.08 – LIHU AREA, WUXI, CHINA" takes a 200,000m² buildable zoning lot (inside the property lot with bigger dimensions) located in the south-east part of Wuxi city, China.

The North border of this land is along the Taihu Lake Avenue, the east is along the Hongqiao road, the south is along the natural river, Dingchang river and the west is along the Taihu Lake.

The area appears to be of exceptional paesistic value because of the condition of a waterfront both west towards Taihu Lake, and south towards Dingchang river and of the view of the mountain chains to the north.

The masterplan presents two main keywords which are the base of all design choices and that can be defined as generator elements of the entire project: landscape and waterscape.

New City Landmark — Mixed-use Architecture

Landscape

The ground is not thought as a flat plate element on which the buildings are simply located, on the contrary it is used as if it would be a "3D solid material" and shaped both horizontally and vertically.

In this way force lines are developed from east to west, always dynamic, which find their highest points on the 2 hills which enclose the villas, that all have a view to the Taihu Lake.

The movement of these force lines produces in-between spaces, which create the squares system and the whole public spaces.

This movement partially generates the buildings themselves, mainly the 5 high rise buildings, the hotel, the facilities and the shopping area, which in this way assume a plastic and dynamic shape.

Other buildings appear to be "on" the landscape instead, particularly the villas, the courtyard building and the multy-storey buildings, these last two detached from the ground, and anyway always using green as a "design material": private gardens and roof gardens for the villas, green facades and roof gardens for the courtyard buildings and hanging gardens for the multy-storey buildings.

Tight linear trees add up to this integrated landscape system following and underlining the landscape bands and represent an integrated project system of essences, most of all made of plum trees.

Waterscape

The intention is to emphasize to the maximum the presence of water making it become a diffuse system, introducing it inside the area and in every part of the project, and using it as if it was a "3D liquid material".

Following as a model the idea of the city of Venice an estuary is created to the south with an artificial island for open air shows, a series of buildings which represent the Yacht Club thought as an archipelago, a small lake system, a timed waterfall (for example 10 minutes per hour, mainly in the evening and with scenographic artificial lighting) coming from the last floor of the hotel.

In this way a perfect fusion between earth, water and architecture is gained in complete harmony of the parts according to a design and attention to landscape approach typically Italian and Mediterranean and pursuing the goal of always researching as a reference the human scale.

The result is a "Landscape Resort Community Park" in respect to the law of humanity and nature.

这是由 ATENASTUDIO 建筑事务所与 Archmaster 工作室合作完成的中国无锡蠡湖区综合规划设计。

该综合规划项目名称为"建筑项目的概念设计，地块 A.02.08——中国无锡蠡湖区"，占地 200,000m²，位于中国无锡的东南部。

规划范围东临鸿桥路，西临太湖，南临鼎昌河，北临太湖大道。

地块西——南面临水，北面靠山，整体自然景观极为出色。

因此，整个 200,000 ㎡ 的建筑规划以"陆上景观"和"水上景观"这两大关键概念为基础进行创意衍生。

陆上景观

地面不是被视为平地，而是可雕琢的"立体的固态材质"，从水平方向和垂直方向进行了雕塑。

通过这种方式，有力的线条从东向西呈动态延伸开来，在包围着别墅的两座山上达到最高点，在那能看到太湖。

产生于这空间中的有力线条的移动，用于营造广场体系和整个公共空间。

移动部分由建筑本身产生，包括五栋高层建筑、酒店、公共设施、购物区，他们通过某种方式形成一个塑形的动态形状。

而其他建筑出现在景观之"上"，特别是别墅、庭院建筑和多层的建筑，后面两种脱离地面，它们都使用绿色作为一种"设计材料"：别墅区的私人花园、屋顶花园，庭院建筑的绿色外立面、屋顶花园，多层建筑的空中花园。

呈线性密集分布的树木添加到这个综合景观系统中，它们遵循并强调景观带，体现出这个综合项目体系的精髓，树木主要是梅树。

水上景观

设计团队尤其注重水的运用，以水为"立体的液态材质"，弥漫于整个建筑规划之中，打造柔性质感。

借鉴了水城威尼斯的表现手法，在南面创造了一个可供露天演出的人工小岛，游艇俱乐部的建筑群则变形为一系列海岛造型，当夜幕降临的时候，从酒店顶层落下的时控瀑布和水幕电影、水幕声光、灯效都极为美妙。

就这样，在参考意大利和地中海特色的景观设计手法和通过不断追求以人为本的的角度为参照物下，泥土、水和建筑完美地融合在了一起。

这儿仿佛是一个尊重人与自然和谐统一的"风景名胜区的社区公园"。

• New City Landmark — Mixed-use Architecture

Architect · B+H Architects
Chief Designer · Olivier Lopion
Design Team · Xiuwen Qiu, Su Jiang, Guanglin Lu, Stephane Lasserre
Client · Shanghai Oriental Pearl (Group) Co., LTD.
Photography · Yijie Hu
Program · Office, Entertainment, Cultural

Shanghai, China

Riverside International Plaza

滨江国际广场

New City Landmark — Mixed-use Architecture

Riverside International Plaza is bordered by the time-honored Yangshupu water plant which has witnessed Shanghai's great changes in the past 100 plus years' from the west, Yangshupu Road from the north and adjacent to the First Phase of Fisherman's Wharf from the east while faces Huangpu River on its south. Proprietors commissioned B+H Architect to make the fullest use of the excellent location advantage to build Riverside International Plaza as a comprehensive plaza which integrates offices, leisure, tourism, culture and entertainment together.

The project concept originates from unique understanding of Shanghai culture and its geographical location. With an inclusive attitude, the design has combined modern buildings and its surrounding historical buildings together and has successfully stricken a balance between modern and traditional elements of the city.

Office building consists of two tower buildings. One is tall and the other is low-rise. The two contrast materials are used on the facade to create a sail-like slash; reminding people of the sailing boat setting off on the riverside. The three items complement

each other to add to the magnificence of the city.

A group of ship-shaped business complex located at the east of the south block is the landmark building of the whole project. The complex, resembling a luxury cruise ship, echoes to the sail design of office tower building. On the west side, the interspersed small-volume buildings are interconnected by myriads of overpasses. These buildings are really in harmony with the Yangshupu water plant buildings standing there.

Seated itself at the god-given position along the riverside, Riverside International Plaza incorporates humanity history, picturesque riverside scenery and other elements together to create a pleasant effect featuring the mutual reaction between scenery and buildings. While trying to gain the broadest view of river scenery, the Plaza itself has become unforgettable scenery.

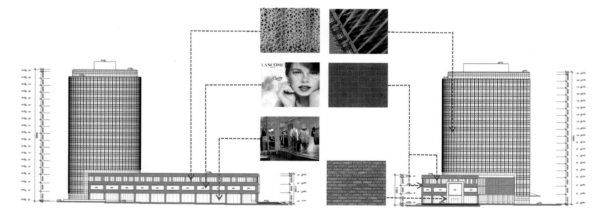

　　滨江国际广场西临见证上海开埠百年历史的老杨树浦水厂，北界杨树浦路，东靠渔人码头一期，南面黄浦江。利用其得天独厚的东外滩核心位置，业主委托 B+H 建筑师事务所将滨江国际广场打造成一个集办公、休闲、观光、文化、娱乐为一体的综合广场。

　　项目构思来源于对海派文化与地块环境的独到解读。设计发挥海纳百川的城市精神，将现代建筑与地块周边相邻的历史建筑有机结合起来，在传承历史建筑脉络的同时，在城市的传统和现代元素之间寻找平衡并建立过渡，将建筑的"现代感"与"历史感"融为一体。

　　办公楼分为一高一低两座塔楼。立面上通过两种不同材质的对比形成风帆型斜线，在与东方渔人码头形态柔和的建筑塔楼相呼应的同时，也让人联想到了浦江边扬帆起航的景象。三者相辅相成，丰富了黄浦江两岸城市天际线的变化。

　　整个项目的标志性建筑当属南地块东侧的一组船形综合商业体，仿佛如停靠在黄浦江边的游船码头的豪华游轮，也呼应了办公塔楼的风帆概念。西侧为错落布置的小体量建筑，通过天桥相连。无论是体量上还是立面的处理手法上，该地块建筑都与西面的百年历史保护建筑杨树浦水厂建筑群形成良好的呼应。

　　坐拥滨江绝版地块，滨江国际广场的设计融合了人文历史、黄浦江美景和亲水绿地等元素，体现了景观与建筑的相互渗透，从而取得相得益彰的效果。在创造最大化的观赏江景视线的同时，又为黄浦江两岸的美景群增添了一道新的亮丽风景。

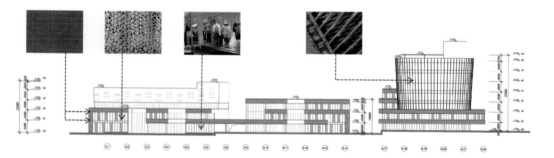

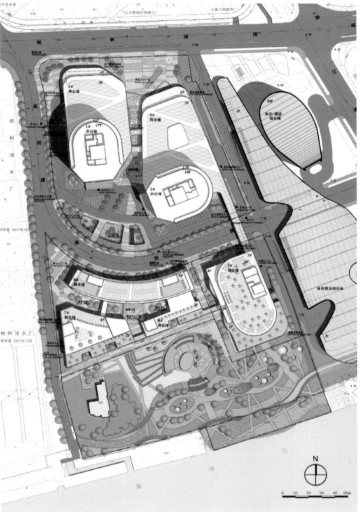

• New City Landmark — Mixed-use Architecture

Architect · B+H Architects
Chief Designer: Jean Sebastien Bourdages
Project Director: Bin Lin
Project Manager: Susan Jiang
Client: Sandhill Equity Ltd., Co
Location: Shanghai, China
Area: 87,338 m²
Photography: Raphael Olivier

Shanghai, China

Sandhill Plaza

展想广场

Located at the core district of Zhangjiang High-Tech Park, Pudong District, Sandhill Plaza takes up an overall building area of 87,388m². The Plaza consists of a 20-storey office tower, eight 3-storey creative industrial R&D buildings and an underground parking lot.

The design focuses on context, axis and courtyard, building up a well-organized and vigorous multi-functional park. The office tower, as high as 100m, stands out as the tallest building in the park. By extending the horizontal lines on the building top, the building looks more magnificent. The panoramic French glass windows lead to a marvelously broad view.

The R&D buildings in the southeast are arranged in an L shape and the business supporting facilities surround an area among the R&D buildings, providing a cool shelter for office staff in the searing hot summer. The exquisitely-

展想广场雄踞浦东张江高科技园区内核心地段，总建筑面积约 87 338m²，由一幢 20 层办公塔楼、八幢 3 层高的创意工业研发楼及一个地下停车库组成。

设计创意以脉络、轴线和庭院为主线，营造了一个有序却不失活力的多功能园区。办公塔楼的建筑高度为 100m，是张江高科技园区内的最高建筑。通过加密建筑顶部的水平线条，在视觉上进一步衬托出建筑的高耸壮观。全景玻璃落地窗的设计，使其拥有非常开阔的视野。

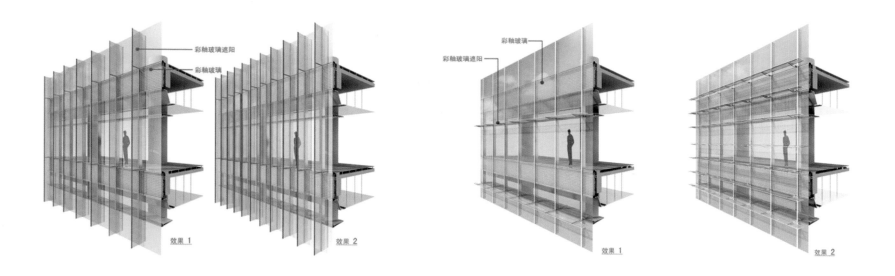

designed lines on "the fifth facade of the building" — the roof not only enriched the colors of the building but also highlighted the uniqueness of it.

Besides, the main courses in the park are in the northwest corner so that people can access to the central green field and get into their office buildings.

The building design concept of green and energy-saving was carried out through the whole designing process, such as supplying hot water by solar heating system and gathering rain water to irrigate green field and sprinkle the road etc. Besides, the layout and building distance arrangement of the park were conducted with factors such

东南边呈"L"型排列的研发楼与商业配套设施围合，形成中心绿地并共享绿地景观，也为办公人员在酷热夏季提供了一个舒适的休憩空间。通过对"第五立面"，即屋顶的独特几何纹路设计，不仅丰富了建筑色彩，更彰显了建筑的独特个性。

此外，进入园区的主要流线设置在园区西北角，以方便人们快速进入中心绿地，继而抵达各自办公楼宇。

绿色节能的建筑设计理念贯穿于项目的整个设计流程中，如利用太阳能热水系统提供生活热水；通过雨水收集回用系统实现景观补水、绿化、浇洒道路等，从而有效利用了水资源。此外，南北朝向与综合考虑日照等因素的平面布局及间距考

as building orientation and sunlight being taken into full consideration so as to maximine day lighting and proper ventilation for the building. Meanwhile, the designers use vertical or horizontal shadding clads, coupled with Low-E glass walls made up of double glazing glasses to preserve heat, thus reducing energy consumption.

Sandhill Plaza meets people's demand for comfort and in the meanwhile effectively reduces energy consumption, yielding winning results in terms of economy and environment. This program was accredited the "one-star level of green building design" certificate by the Ministry of Housing and Urban-Rural Development of the People's Republic of China.

虑,保证了建筑能获得足够的采光及有效的通风。与此同时,根据不同的朝向,我们设置水平或是垂直的遮阳线条,加上双层中空 Low-E 玻璃幕墙,以达到保温隔热,抑制能源消耗的目的。

　　展想广场既满足了人们对舒适性的需求,同时又有效地降低了建筑能耗,达到了舒适性、经济性双赢的效果,提供了良好的社会效益。本项目被中华人民共和国住房和城乡建设部授予"一星级绿色建筑设计标识"证书。

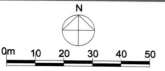

- New City Landmark — Mixed-use Architecture

Architect · F. Bozkurt Gürsoytrak
BOYUT ARCHITECTURE CO., LTD. &
Norman Foster FOSTER + PARTNERS
Client · Capital Partners
Area · 339,344 m²
Program · Entertainment, Retail, Cultural, Office, Residential

Almaty, Kazakhstan

Almaty Financial District North Phase Complex

阿拉木图金融区

The aim of the project is redefine the city of Almaty as the financial and cultural heart of Central Asia and this is the fourth phase complex which is situated to the north of the city's major access route, Al-Farabi Avenue.

The twin towers are positioned to address the center of the district, and will act as a landmark for the surrounding area. They will be the tallest building in Kazakhstan when completed.

The project mainly consists of two 216m tall, 48-storey office towers which are linked at the lower levels, and entered through a dramatic shared atrium space. And also they will be flanked by smaller towers rising from 8 to 15 floors towards the twin towers. These gradually rising towers planned as 120 apartments of various sizes and retail spaces at ground levels are flexible with different sized units.

All these 4 towers create an enclosed central plaza with shops, cafes, public space and cultural facilities – such as the amphitheater in the core of the plaza – the building will create a vibrant and lively environment.

The three combined components of the scheme will be served by a shared underground car park offering approximately 1,600 parking spaces, distributed over three underground levels. Basement level1 will be served for residents and retail customers, with level 2 and 3 available for office workers.

Towers are standing roughly elliptical in shape with points at the ends of their main axises, having glass facades with the diagrid pattern. The glass facade will give people

New City Landmark — Mixed-use Architecture

spectacular views of the surrounding area, maximizes the views to the mountains and at the top of the towers will be several levels of roof gardens where people can go to relax.

The two residential buildings are situated at the perimeter of the central plaza, and follow its circular shape. Their configuration functions as a protective barrier, which blocks unwanted noise and views from the highways. Their sloping shape emphasizes the importance of the towers while at the same time provides luxury apartments with roof gardens and unobstructed views towards the mountains while fully exploiting the environmental potential of the site.

该项目的目标是重新定义阿拉木图为中亚的金融和文化中心，这是综合体的第四期，位于城市的主要通路 Al-Farabi 大道上。

双子塔的位置是区域的中心，它们将作为周边区域的一个地标，也将成为哈萨克斯坦的最高建筑。

项目主要包含两座 216m 高 48 层的办公塔楼，它们在较低层处相连，通过一个引人注目的中庭空间进入。它们的侧边是 8 层和 15 层的小塔楼。这些朝向双子塔、逐渐上升的小塔楼用作不同大小的零售空间，共 120 室。在底层，不同大小的单元是灵活布局的。

所有 4 座塔楼与商店、咖啡馆、公共空间和文化设施共同创建了一个封闭的中央广场，如在广场中央的圆形露天剧场，建筑将创建

一个充满活力的、活泼的环境。

方案的三个合并的组成部分将共享一个地下停车场，该停车场大约1 600个停车位，分布在三个地下层上。地下一层供居民和零售顾客使用，地下二层和三层供办公室工作人员使用。

竖立的塔楼类似椭圆形，在他们的主轴线上有端点，玻璃外立面呈斜肋构架模式。玻璃外立面使人们将周围的壮观景致尽收眼底，以及最大程度地看到山脉。塔楼地顶部将是几层屋顶花园，人们可以在那里放松、休闲。

这两个住宅坐落在中心广场的周边，并绕其圆形而建。它们的配置功能作为一个保护屏障，阻挡不需要的噪音和高速公路上的川流不息的车流画面。它们倾斜的外形强调塔楼的重要性，同时提供豪华的屋顶花园公寓，在公寓里朝向山脉的视线畅通无阻，这充分利用了场址潜在的环境优势。

- New City Landmark — Mixed-use Architecture

Architect · F. Bozkurt Gürsoytrak BOYUT ARCHITECTURE CO., LTD. & Paul Molle Design International
Client · Ilhan Cavcav, Ankara Flour Mills Inc.
Area · 231,703 m²
Program · Retail, Cultural, Entertainment, Office, Residential, Hotel, Restaurant

Ankara, Turkey

Cavcav Center

卡维卡维中心

Cavcav Center is located on Konya Road which serves as the main city transportation axis that connects city center to the airport. This area is considered to be one of preliminary commercial centers in terms of real estate sector.

Cavcav Ankara project aims to be a small model of the city with its mixed use complex design. The values lost in the city of Ankara, are given place in this project as much as possible.

卡维卡维中心位于康亚路上,是这座城市主要交通轴,连通着城市中心和机场。该区域是房地产区域准商业中心之一。

拥有多功能设计的卡维卡维·安卡拉工程旨在成为这座城市的一个小型典范。安卡拉城市价值的降低,尽可能地为这项工程提供可能。

商场区域设建筑面积86 894m²,设有儿童区域、家庭娱乐区域、超市、电子市场、零售商店、咖啡厅,酒吧、饭店多功能大厅。

Mall area contains children and family entertainment areas, supermarket, electronic market, retails, café, bar, restaurants and multi purposed hall within its 86,894 m² construction site.

Whole complex is supported by 3-storey parking area under the building with its designated car capacity of 4,472 cars and 111,029 m² construction area. Complex also contains a 14-storey office-residential block and a 17-storey hotel block. Office block

整个复合式建筑由一个3层的地下停车场支撑,该停车场能容纳4 472辆汽车,建筑面积111 029m²。此复式建筑同时包含了一栋14层的办公居住两用楼,1栋17层的酒店。一流的办公楼总共建筑面积为12 018m²。酒店建筑面积为116 113m²,是一家4星级商业连锁酒店。

卡维卡维中心是一个居住楼,24小时不熄灯。卡维卡维中心及其酒店、办公楼、公寓、电影院、剧院、多功能大厅成为了安卡拉街道的代表。与安卡拉其他类似的零售中心不同的是

New City Landmark — Mixed-use Architecture

is planned to be "A class" and has a 12,018 m² construction area in total. Hotel block with 116,113 m² construction area is planned to be marketed as a 4-star business-hotel chain.

卡维卡维中心既有封闭的空间亦有开放的空间，不同的高度给人们带来的是不同的景色，不同的视野，不同的角度，让人们身临其境的感受其周边环境。此外，卡维卡维中心的剧院和表演

Cavcav Center is designed to be a living building with its lights never turned off whole day. It is planned to represent the streets of Ankara, with its hotel, bazaar, offices, apartments, cinemas, theatres and multipurpose hall. Unlike other similar retail centers

中心将城市文化发扬光大。

一般来讲购物中心的屋顶常常是巨大的无用空间，这项工程中，周边风景极大地丰富了

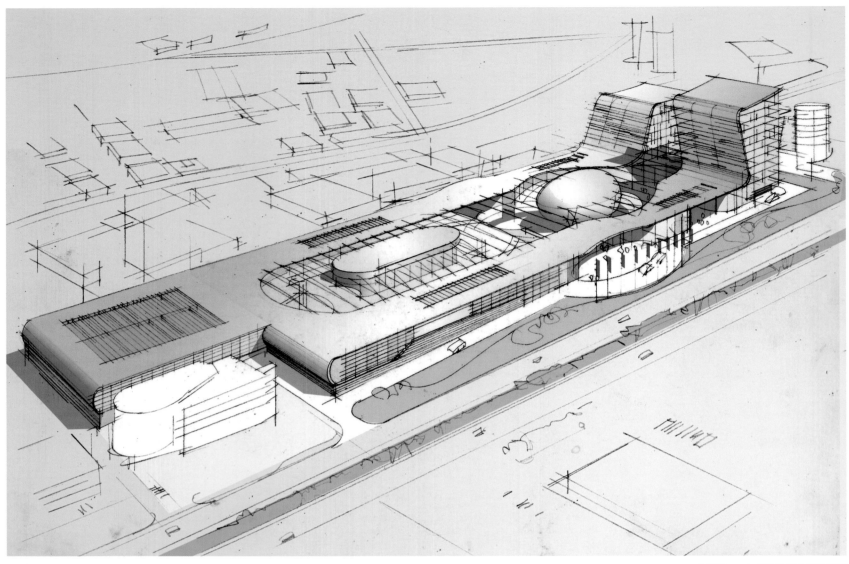
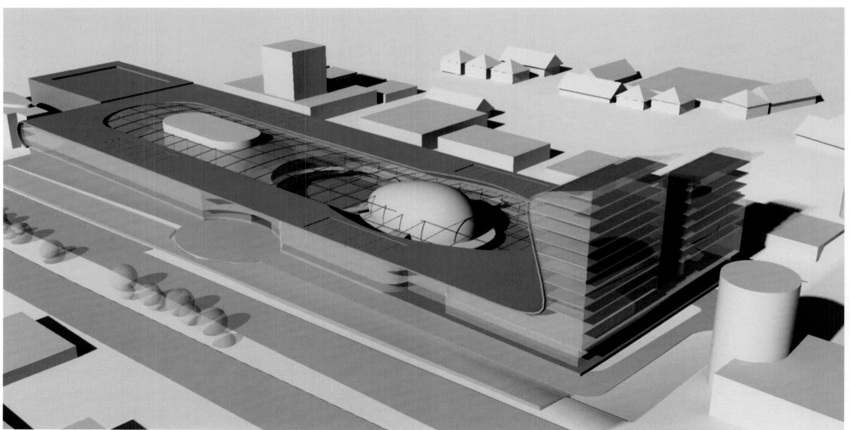

in Ankara, there are closed and open spaces, with different heights to create different perspectives, exposure points, angles which provide people to feel the surrounding. In addition, it is aimed to make contribution to the city culture with its theatre and performance center.

Always there is a very large roof area unused space in shopping centers, at this project the roof area is enriched by landscaping and having green roof. It helps to reduce the urban heat island effect, as well as creating a new park on the roof to contribute to a sustainable city.

Instead of closed boxes that have no connection with the city, a transparent building which is in contact with the surrounded streets is designed. Building tries to speak to the city with the help of its connections with its own inner space and outer space.

Building gathers different facilities like high-rise hotel and

office blocks, shopping malls that are divided into two by the concept of contrast; old-new, modern-traditional, open-close, east-west; all are under one belt structure.

Main plaza where exhibitions, shows are performed and people meet locates in the middle of two conceptually differentiated malls and distributes people to the other spaces. Transparent egg shell formed entertainment center in the middle of the new, modern and open mall on the left side is active for 24 hours and it creates a living space around the street / outdoor shopping area.

屋顶区域，因此该项目有了绿色屋顶。绿色屋顶可以帮助减少城市的热岛，同时也在屋顶上建立了一个新的公园，为可持续发展城市作出积极贡献。

卡维卡维没有设计成与城市毫不相通的封闭式建筑盒子，而是设计成了透明的建筑，与周围街景的设计形成了鲜明对比。该建筑通过它里外空间的相连努力地与这座城市进行亲切对话。

像高层酒店，办公楼，购物商场等各式各样的建筑部分统统集中在这座建筑物中，根据不同的概念他们被分为两组；新或旧、现代或传统、开放式或封闭式、朝东或朝西，所有这些都在单区域结构下。

主广场位于两个不同概念的购物中心之间，人们在此可观看展览和演出，在此聚集并由此前往他处。透明的卵壳就是娱乐中心，位于崭新的现代感十足的开放性商场中央，其左侧 24 小时开放，为街道和户外购物区域的周边创建生活空间。

• New City Landmark — Mixed-use Architecture

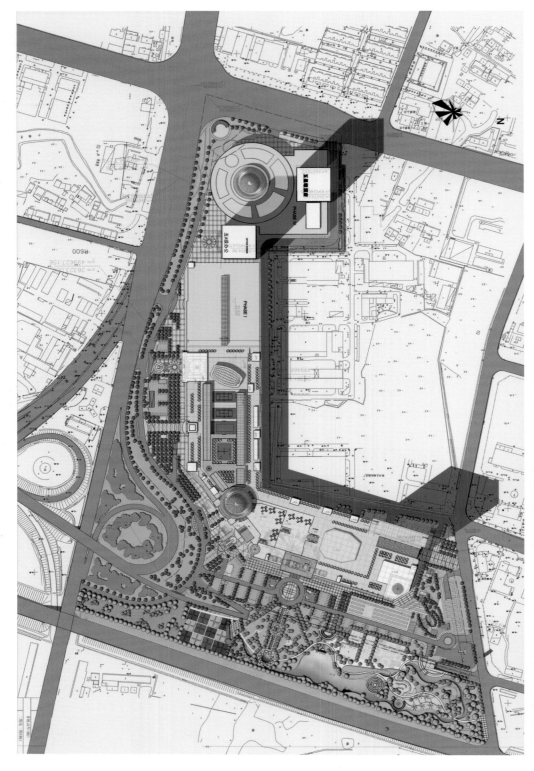

Changzhou, China

Global Harbor, Changzhou

常州环球港

Architect · Chapman Taylor
Design Team · Chris Lanksbury, Hualei, Doreen Wang, Rodney Carran, Peter Mackey, Jason Lei, Able Wang
Area · 878,900 m²
Program · Retail, Office, Restaurant, Entertainment, Cultural

New City Landmark — Mixed-use Architecture

Global Harbor, Changzhou, is one of the super-large mixed-use complexes in Changzhou City, even in the region of Jiangsu, Zhejiang and Shanghai. The finished building will become a mark of Changzhou, especially its sitting in the entrance of highway which makes it more attractive. This commercial center mainly locates a general commodity supermarket and a special commodity supermarket, bringing together restaurants, offices, leisure, culture, recreation and gym.

Global Harbor is a complex business development program, with a narrow "L" type base, total length of 700m from north to south, gross breadth of 213m, and the narrowest area of a 117m breadth. According to the topography, building groups are arranged in a "L" shape. And two pedestrian malls with circle passages and varied spaces connect main stores at both ends of this building group. With a high flexibility, just like a agile dragon, the two pedestrian malls lie on the north gate of Changzhou. And a giant lobby with a dome artfully links the commercial area on both sides. The whole design of the building group is filled with visual tension, endeavoring to create an elegant and comfortable super shopping and leisure places.

From the second floor to the top floor, the forth floor, the three floors gradually retreat, provide an opening outdoor "Carnival". Its function emphasizes on entertainment and outdoor restaurants accompanied by rooftop Ferris Wheel. Roof gardens of each floor offer us both indoor entrance and outdoor entrance. And footsteps bring people from a outdoor square on the bottom floor to the second floor, then the third floor and the forth floor. The interesting special organization fully capitalizes on rooftop gardens to provide leisure function.

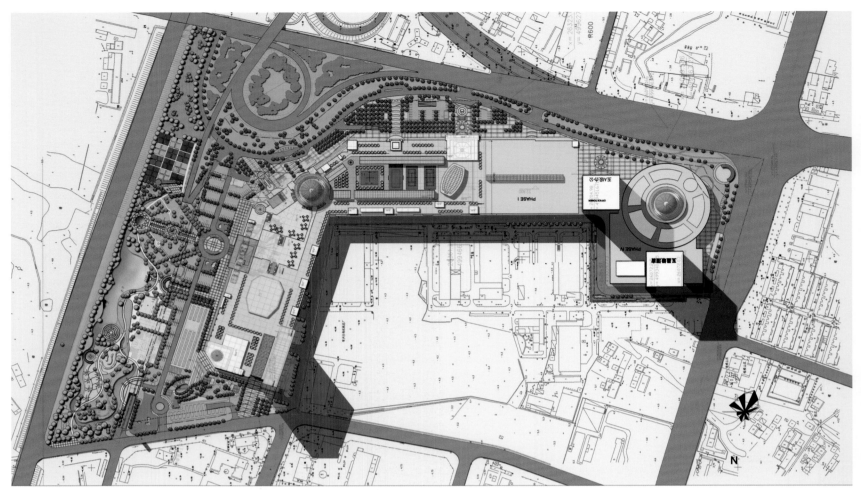
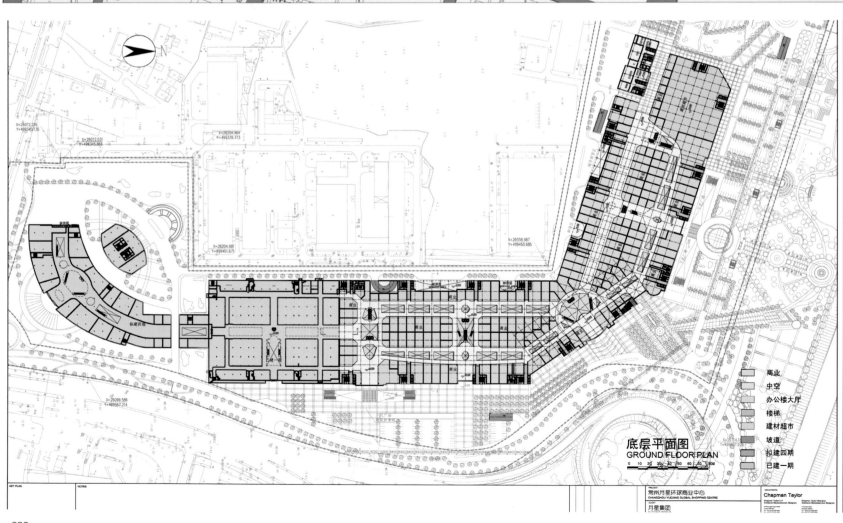

底层平面图
GROUND FLOOR PLAN

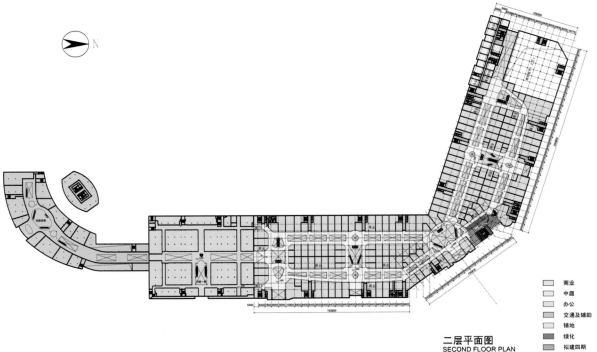

二层平面图
SECOND FLOOR PLAN

商业
中庭
办公
交通及辅助
铺地
绿化
拟建四期
已建一期

Interior design of Global Harbor is also outstanding. Traditional Europe style serving as a general background of interior space, the whole interior space is divided into three Courts and two Commercial Porches. The three courts fall into three types: "Garden Court", "Rotunda" and "New York Plaza". The two porches are "Garden Avenue" and "Mediterranean Arcade". Garden Court and Garden Avenue are a garden area that advocates the conception of environmental conservation. More expansively and gloriously, there is a super 6-floor indoor garden with

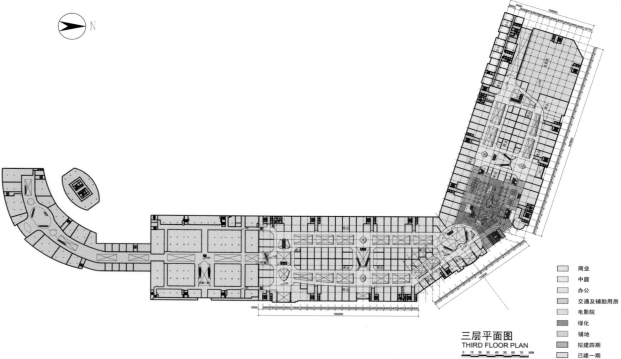

三层平面图
THIRD FLOOR PLAN

商业
中庭
办公
交通及辅助用房
电影院
绿化
铺地
拟建四期
已建一期

a void structure from B1 to L5, offering an important commercial sight of relaxing shopping, fashion consumption, and sightseeing. Compared with garden area, "Rotunda" and "Mediterranean Arcade" are more magnificent and spacious. Classical Europe court with red as main color and very high cross-layer pillars is extremely impressive and luxurious. New York Plaza, which is more fashion and modern, provides an indoor New York Fashion Square. There are a huge sculpture, relaxing and fashion water landscape, and a large stage for activities, creating

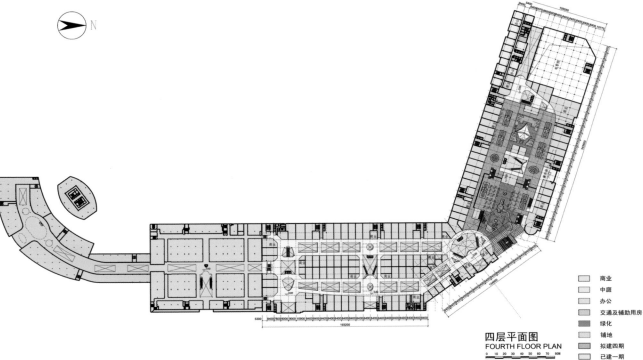

N

四层平面图
FOURTH FLOOR PLAN

商业
中庭
办公
交通及辅助用房
绿化
铺地
拟建四期
已建一期

a lively and bustling recreational commercial space.

The major designer employed his rich experience for commercial design to conceive out this internal topped and luxurious commercial center align with the grand and splendid plan. Global Harbor definitely is invaluable in both business value and art value.

常州环球港是常州市，同时也是江浙沪一带超大型的综合商业项目之一。整体建筑建成后将成为常州市的标志性建筑，尤其是它的地理位置在高速路的入口处会更加吸引人。该商业中心将是一个以大型综合百货和专项百货为主，集餐饮、办公、休闲、文化、娱乐、体育健身为一体的休闲中心。

常州环球港是一座综合型商业开发项目，基地呈狭长的"L"型，南北总长约700m，东西最宽处约213m，最窄处约117m，本设计将建筑群结合地形布置成"L"型，其中用两条

空间富于变化的环形通道商业步行街连接建筑群两端布置的主力店。这两条灵动的步行街就犹如摆尾的神龙，横卧在常州市的北大门上，当中一个大型的圆形穹顶大厅把两侧的商业巧妙而有机地连接在一起。整个建筑群体设计充满了视觉的张力，力图创造一处格调高雅、环境舒适的现代城市高级购物、休闲场所。

本项目的二到四层屋顶层层退台，形成对外开放的室外"嘉年华"。其功能着重于娱乐和室外餐饮，并设有屋顶摩天轮。各层屋顶花园均可从室外和室内直接通达，室外以台阶方式把人从底层室外广场引至二层屋面，再上至三层及四层。以饶有趣味的空间组织充分利用屋顶花园，发挥其休闲娱乐的功能。

常州环球港的室内设计也尤为精彩。其室内空间以经典的传统欧式风格为大框架背景，把整个室内空间划分成三个中庭观光区和两个商业长廊。三个大中庭分为："Garden Court"（花园中庭）、"Rotunda"（天穹大厅）和"New York Plaza"（纽约广场）区域。商业长廊为："Garden Avenue"（花园大道）和"Mediterranean Arcade"（地中海走廊）。其中花园中庭和花园大道是以倡导绿色环保理念的花园区域。设计更为宽阔、壮丽，从B1直到L5的6个六层大面积挑空的室内超级大花园。从而形成一个轻松购物、时尚消费、旅游观

光的重要商业风景线。和花园区域相比，"天穹大厅"和"地中海走廊"区域则更为壮观、宏伟。红色基调加上高大跨层的柱子打造出来的经典欧式中庭叫人叹为观止，尽显大气奢华。纽约广场更显时尚现代风格，打造室内的纽约时尚广场。中庭有26m高的大型雕塑、休闲时尚的观光水景和活动必备的大型舞台。打造热闹，沸腾的娱乐商业空间。

主设计师运用多年的商业设计经验融合宏伟、壮丽的设计打造出一个国内顶级的、奢华的商业中心。常州环球港无论在商业价值上还是在艺术价值上都是无可限量的。

• New City Landmark — Mixed-use Architecture

Tangier, Morocco

Mall of Tangier

丹吉尔购物中心

Architect · Design International
Project Principals · JDavide Padoa (CEO), Lucio Guerra (Managing Director), Paul Mollé (Founding Partner)
Client · Groupe Aksal
Area · 33,150 m²
Program · Retail, Entertainment, Restaurant

New City Landmark — Mixed-use Architecture

Mall of Tangier is located on the most prominent intersection between Avenue Mohamed VI and two new marinas, which are part of the ongoing transformation of the port of Tangier into a commercial and cultural hub, with hotels, a palace of congress, a museum, residences and Mall of Tangier as the new leisure and shopping destination of the city. Given the great exposure of the project, Mall of Tangier has been designed as a sculpted diamond that defines with immaculate precision its corners and facades. The faceted facades of the project, which mimic an inverted pyramid that has been plugged into the ground, reflect the sea on one side and the structure of the old médina on the other side of the city. This sculptural shape is further enhanced by the ambition of the project to create a new recreational focal point in the city through its indoor facilities as well as a landscaped outdoor area, which extends the surrounding park through a green ramp to the roof of the building.

The concept for Mall of Tangier aims to maximise the transparency of the mall and to create the first "5D shopping and leisure experience in Africa" with a building that visitors can "touch, feel, walk, discover and climb - all at the same time", while the roof area will offer dining experience with the most stunning view in the whole of Tangier.

Inside the transparency between each one of the three retail and leisure floors as well as a great amount of beautifully landscaped green areas will become the main attraction. Mall of Tangier will feature a luxury plaza with all the best brands, a 4,000m^2 gourmet hypermarket, an Imax theatre a traditional souk with prayer room, a fun park, a fitness club and more than 100 retail, cafe and restaurant units. Mall of Tangier will feature also the tallest climbing rock and waterfall ever built in a mall.

丹吉尔购物中心位于穆罕穆德六号大街和两个新码头的突出十字路口上，作为丹吉尔港文化 — 商业化转型的一部分，其将与酒店，国会宫，博物馆，住宅一起，成为新的城市休闲和购物胜地。鉴于项目的巨大风险，丹吉尔购物中心的任何边角和门面都犹如钻石一样精雕细琢。项目雕琢平面的外立面，仿佛一个倒置的金字塔插入地面，一面映射着大海，一面映射着城市另一面的老城风景。此雕琢的形状将会得到更新，以其室内设施以及室外景物，将该项目打造成为城市新娱乐休闲中心，该项目的室外景物，将会通过绿色斜坡，将公园一直延伸至屋顶。

丹吉尔购物中心的概念旨在最大限度地提高商场透明度，在非洲营造首个5D购物和休闲体验场所，实现参观者一站式集合触摸、感觉、行走、发现与攀爬的建筑，而屋顶区域将会让游客在丹吉尔最壮丽景色下享受就餐体验。

三个零售区之间的各个高透明区域和休闲地板，以及大量的美丽景观将会成为主要吸引点。丹吉尔购物中心将会营造一个给所有大品牌的奢华广场，一个4 000m²的美食超市、一个IMAX影院、一个带祈祷室的传统露天市场、一个趣味性的公园，一个健身俱乐部和超过100多家零售、咖啡厅和餐馆。丹吉尔将成为有史以来在购物中心里面建造最高的攀岩和瀑布的购物中心。

- New City Landmark — Mixed-use Architecture

Isfahan, Iran

Pardis — Health & Leisure Centre

Pardis 保健休闲中心

Architect · Design International
Project Principals · Davide Padoa (CEO), Lucio Guerra (Managing Director), Paul Mollé (Founding Partner)
Client · Hermes
Area · 85,281 m²
Program · Conference, Hotel, Retail, Entertainment, Public Facilities

New City Landmark — Mixed-use Architecture

Pardis Health & Leisure Centre is located in Isfahan, Iran's third largest city with 3 million inhabitants, which is about 450 km south of Teheran. Pardis is a true mixed-use development, consisting of a Medical Centre with a whole range of private medical facilities, a Conference Centre, a Hotel & Spa, a Shopping Centre and two large outdoor gardens.

The architecture of the development takes inspiration from the style and design of the traditional Persian Garden, or "Paradise Garden", which has influenced the design of gardens all over the world. The typical Paradise Garden features an enclosed wall, rectangular pools, and internal network of canals, garden pavilions and lush planting. The concept for Pardis is a contemporary interpretation of this ancient Paradise Garden, which presents itself as two distorted rectangular gardens, embraced and interconnected by the main components of the project. The main architectural feature of the development is an iconic 17-storey leaning tower, which houses the main bulk of the medical facilities and an underground conference centre underneath.

Strong use of local materials such as stone and copper on the exterior elevation of the building reflect the surrounding earthly qualities and is seen as a shield against the elements. Whilst in contrast a deliberately light, transparent and open facade is used on the inner elevations overlooking the garden spaces, creating the same "introverted" atmosphere so common to Iranian urban design.

Pardis Health & Leisure centre, through a mix of world-class medical services, luxury hotel apartments and upscale shopping & entertainment offers will cater for the wealthier population of Isfahan, the wider area, as well as national and international business people and tourists coming to the area.

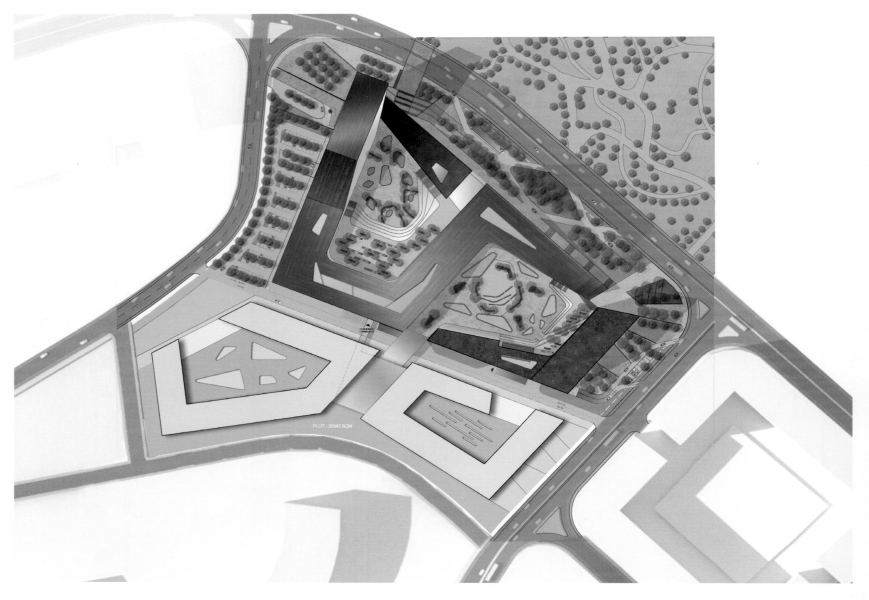

Pardis 保健休闲中心坐落在伊斯法罕,伊斯法罕是伊朗的第三大城市,拥有三百万人口,在德黑兰以南大约 450km 处。Pardis 是一个真正的综合体,包含拥有一系列私人医疗设备的医疗中心、会议中心、酒店及水疗中心、购物中心和两个大型户外花园。

建筑的开发从影响了世界各地花园设计的传统波斯花园或"天堂花园"的风格与设计中吸取灵感。典型的天堂花园以封闭的墙、矩形池和内部网状分布运河、花园凉亭和郁郁葱葱的植被为特色。Pardis 理念是对古天堂花园的当代诠释,它表现为两个扭曲的矩形花园,花园

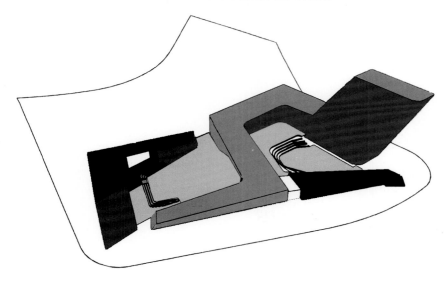
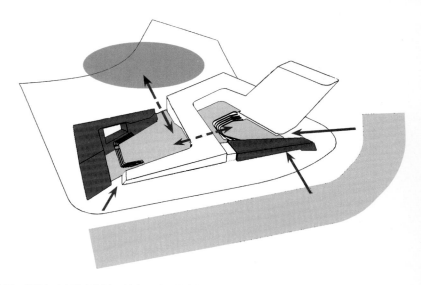

被这个项目的主要元素环绕和相互连接。这个开发体的主要建筑特色是一个标志性的 17 层的斜塔,斜塔内是大部分的医疗设施和下面的一个地下会议中心。

大量使用当地材料,如在建筑外立面的石材、铜,这反映了周围环境的品质,是对建筑元素的一种保护。相比之下,再俯视花园空间的内侧,刻意使用的一个轻的、透明的开放式外立面,营造与大部伊斯法罕城市设计一样含蓄的氛围。

Pardis 保健休闲中心,将通过世界级的医疗服务、奢华酒店公寓、高档的购物和娱乐,为伊斯法罕地区大量的人口以及国内、国际游客提供服务。

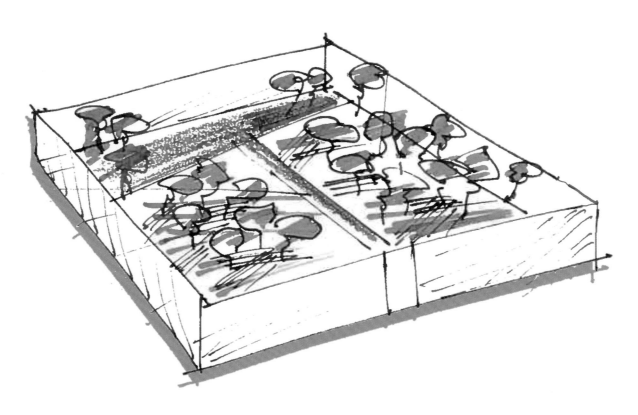

• New City Landmark — Mixed-use Architecture

Marghera, Italy

Romea

罗米亚

Architect · Design International
Project Principals · Davide Padoa (CEO), Lucio Guerra (Managing Director), Paul Mollé (Founding Partner)
Client · Corio Italia S.r.l. (owner) / B.L.O. S.r.l. (developer)
Area · 55,600 m²
Program · Retail, Entertainment, Restaurant

New City Landmark — Mixed-use Architecture

Romea shopping centre is located in Marghera, Italy, adjacent to Mestre, which is the landside town that links to Venice via a causeway. The catchment area has a population of over 1million people and has one of the highest spending powers in Italy. In addition, Romea shopping centre will benefit from the huge number of tourists that come to Venice and the area each year. Romea shopping centre, which is currently under construction and expected to open in March 2014, has a total GLA of 40,000m^2, over 110 shops, including a large hypermarket, foodcourt and restaurants, as well as leisure and entertainment facilities.

The architectural concept for the project is inspired by the local area and culture, in particular the merchant trading empire of the republic of Venice, which was a major maritime power at the time. This is reflected in the architectural language and the use of materials associated with ships, most outspoken in the mall entrance and "La Nave", an interior plaza in the shape of a sailing ship. "La Nave" is one of three compact malls, which forms the major components of the interior layout, which is created like a series of interlinked plazas.

The lighting design played a major role in enhancing the open and transparent character of the building. An innovative roof design allows continuous views of the sky and day lighting to the Public areas. Light sculpts every corner within the Romea project, while during the dark hours energy efficient lighting of the highest specification throughout the project is controlled by a central management system to provide an ever changing scenery.

The result is an outstanding architecture of the building, which is matched only by the projects approach to sustainability, which is reflected in the fact that it is on track to receive the first "very good" rating in its BREAM certification in the category of "new built".

  ROMEA 购物中心位于意大利马格拉，毗邻通过长堤与威尼斯相连的路侧镇梅斯特。集水区超过 100 万人口，为意大利最具消费能力地区之一。另外 ROMEA 购物中心每年将会从威尼斯及其附近区域的巨大旅游人数而获益。正在建设当中的 ROMEA 购物中心，预计在 2014 年三月份开业，拥有总可出租面积 40 000m^2，超过 110 多家门店，包括大卖场、美食广场、餐厅以及休闲和娱乐设施。

  该项目建筑灵感来自于当地地区及其文化。尤其是作为曾经主要海上力量的威尼斯共和国的商业贸易帝国。这些文化反映在建筑语言以及与船舶有关的建筑材料上，在最显眼的购物中心入口和 LA NAVE 上，LA NAVE 是一个在航海船舶形状建筑内的室内广场。LA NAVE 是三个小型商场之一，其构成一系广场内部相连的室内布局主要组成元素。

  灯光设计在提高建筑开放性和透明度上扮演者主要的角色。创造性的屋顶设计让人们在公共区域上可以看到连续不断的天空景色，享受直射阳光。在 ROMEA 项目内，阳光照遍每一个角落，在夜间，贯穿整个项目的最高配置的高效节能灯由一个中央管理系统控制，塑造不停变化的美丽景色。

  该项目将成为杰出的建筑，追求可持续发展，折射出其正在获得新建筑群 BREAM 证书的第一个"优秀"道路上的事实。

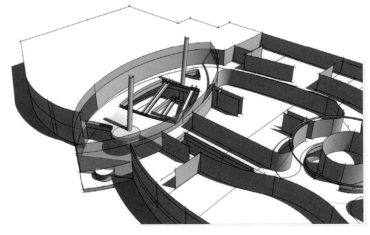

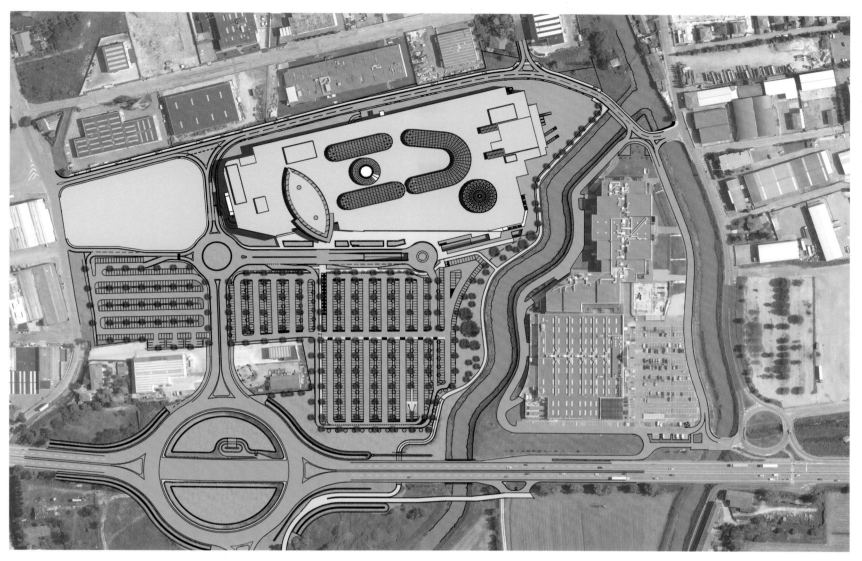
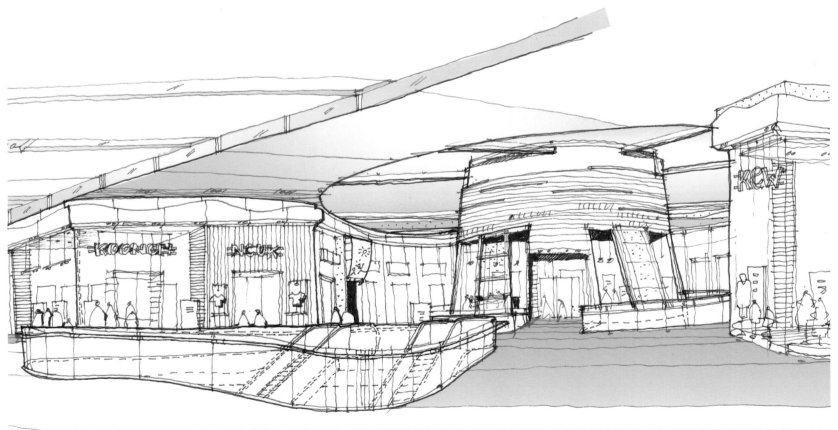

- New City Landmark — Mixed-use Architecture

New City Landmark — Mixed-use Architecture

Architect · Design International
Project Principals · Davide Padoa (CEO), Lucio Guerra (Managing Director), Paul Mollé (Founding Partner)
Client · Resiliance
Area · 103,643 m²
Program · Retail, Entertainment, Residential, Office

Marseille, France

TivoliParc

TivoliParc

TivoliParc is a mixed-use development, currently under construction in Marseille, France. The project will include retail & entertainment components, residential units, offices as well as a publicly accessible external landscape.

TivoliParc is developed in conjunction with the refurbishment of Marseille Grand Littoral (Developer: Corio), an existing 120,000m² shopping centre. This joint project is currently Marseille's biggest private regeneration projects in an area of the city, which was in dire need of improved commercial, recreational and business facilities. The project will also help to integrate the area better into the commercial and social activities of Marseille.

Working with the city council of Marseille, TivoliParc & MGL will become the new heart of this urban zone with improved housing as part of Plan d'Aou, which will completely transform this old industrial area of the city.

Design International, which had been working on MGL since 2010, was hired in 2011 for TivoliParc and created a commercial and architectural concept, whereby the two projects integrate and complement each other, rather than compete. The open and transparent architecture of TivoliParc, with a particular focus on the external landscape, creates a balance to the interior focused environment of MGL.

The TivoliParc & MGL complex will provide what had been promised to the population of Marseille a long time ago: a project of true European scale, a place for all their shopping and entertainment needs – all in one place – without having to leave the city.

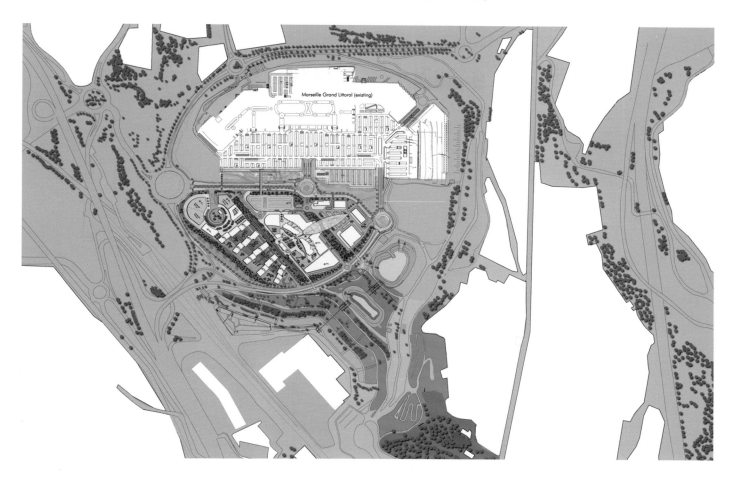

TivoliParc 是位于法国马赛的综合体项目，目前正在建设中。项目包括零售和娱乐、住宅单元、办公楼以及对外开放的外部景观。

TivoliParc 的开发联合了马赛大滨海的翻新，马赛大滨海是一个现存的 120 000m² 的购物中心。这个联合工程目前是其所在区内最大的私人重建项目，这个区急需改进商业、休闲和经济设施。这个项目也有助于该区更好的融入到马赛的商业和社会活动中。

TivoliParc & MGL 与马赛的市议会一起作为 d'Aou 计划的一部分，将改善城市设施，成为这个城区的新中心，这将完全改变这个老工业区。

自 2010 年以来一直从事 MGL 建设工作的国际设计公司被 TivoliParc 聘请来创建一个商业建筑理念，由此两个项目彼此融合与互补，而不是竞争。

TivoliParc & MGL 综合体为马赛的人们提供承诺已久的一个真正符合欧洲标准、满足他们购物和娱乐需求的地方。由此，在马赛人们在一个地方就能获得他们所需的，无需离开这个城市。

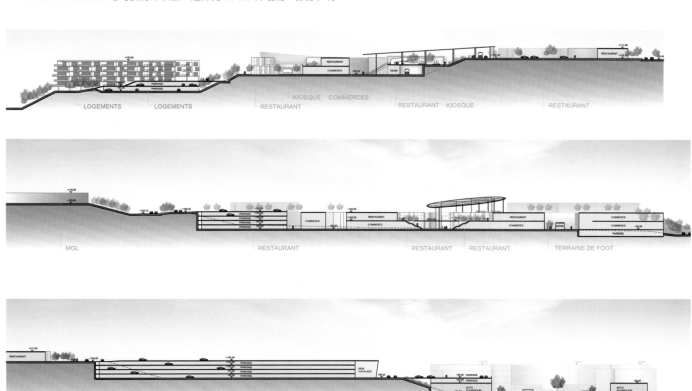

- New City Landmark — Mixed-use Architecture

New City Landmark — Mixed-use Architecture

Architect · Design International
Project Principals · Davide Padoa (CEO), Lucio Guerra (Managing Director), Paul Mollé (Founding Partner)
Client · Qinhuai District Government Nanjing
Area · 270,000 m²

Nanjing, China

Zijin Skywalk

紫金天空步道

The district of Qinhuai has a rich historical and cultural background, and is famous for the beauty and calming atmosphere of its public spaces, which have been inspiration for many famous Chinese poets throughout history. The poetry spread widely throughout China, thanks to an advanced engraving printing technique of that era: the wooden movable type printing, a technique that has been awarded as "Intangible Cultural Heritage" by UNESCO.

This traditional printing technique was the inspiration for the texture and character of the facade of the Zijin Skywalk development, making Nanjing's historical and cultural background an important aspect of the architectural language of the project.

The interior spaces of the project have a wide range spatial quality, with good internal connections, are full of natural light and permanent views to the internal gardens.

These spaces are designed for a variety of uses, ensuring the building can adapt to the needs of the users and therefore achieves greater space occupation, whether it be used for office, research or high tech manufacturing.

Open plan office layouts encourage good communication and provide flexibility to accommodate changing work procedures.

秦淮区拥有深厚的历史和文化背景，尤以优美宁静的公共氛围闻名，成为历史上众多文人墨客吟诗咏赋的灵感来源。而这些好诗佳句得以传颂全国，还得归功于那个时代一种先进的雕版印刷技术——活字印刷术，它已被联合国教科文组织列入非物质文化遗产名录。

这一传统的印刷技术也为紫金天空步道项目设计带来灵感，使南京的历史和文化背景成为项目建筑语言的一个重要方面。

项目的室内空间拥有宽领域的空间品质和良好的内部连通，室内花园更能提供满满的自

然采光和优美景致。

空间设计力求多功能化,确保建筑物能满足用户的不同需求,无论是办公、科研还是高科技制造,都能实现空间的最大化利用。

开放式的办公室格局为良好的人际沟通提供便利,灵活性的设计也能适应不断变化的工作程序。

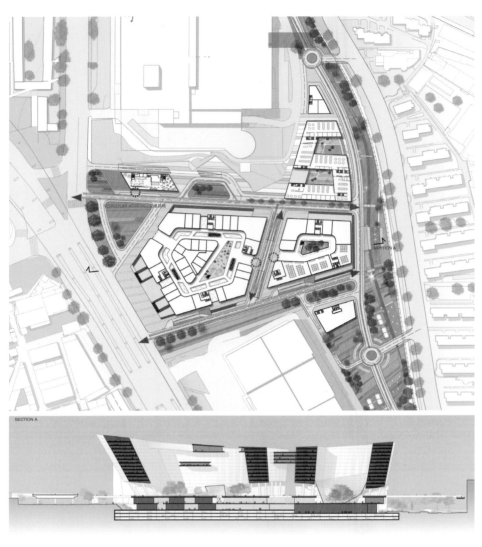
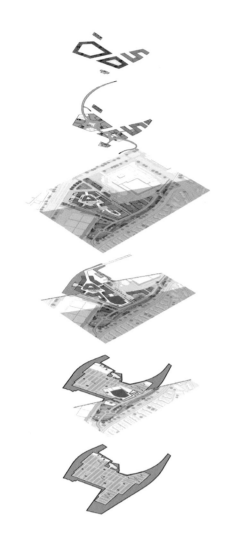

SECTION A

Masterplan concept

Masterplan Form Finding

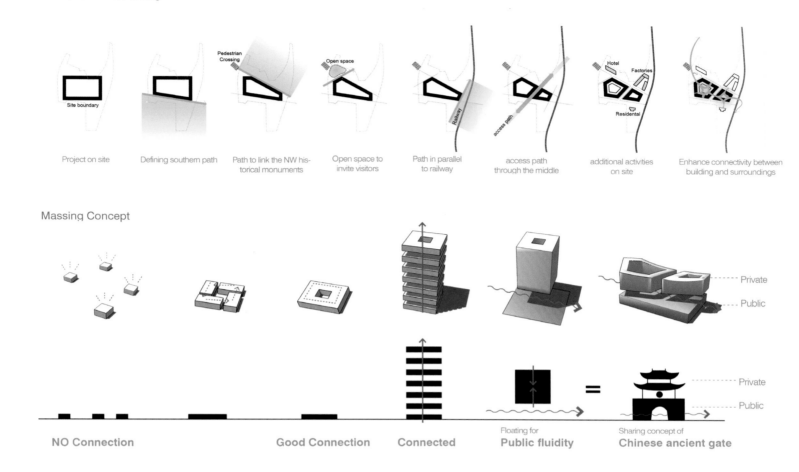

Project on site	Defining southern path	Path to link the NW historical monuments	Open space to invite visitors	Path in parallel to railway	access path through the middle	additional activities on site	Enhance connectivity between building and surroundings

Massing Concept

NO Connection — **Good Connection** — **Connected** — Floating for **Public fluidity** — Sharing concept of **Chinese ancient gate**

01. South Elevation

• New City Landmark — Mixed-use Architecture

Architect · Ector Hoogstad Architecten

Shanghai, China

Zhangjiang Incubator Competition

张江企业孵化大楼

Together with reknowned Design Institute SIADR (Shanghai Institute of Architectural Design & Research) from Shanghai, EHA (Ector Hoogstad Architecten) was one of the competitors in an invited competition for an incubator building in the new urban district of Zhangjiang in Shanghai. The 80m tall building constitutes the heart of a new hightech campus, almost exactly between the city center and Pudong airport. The design forges a strong connection between different programmes such as offices, exhibition space and retail, and sets out to encourage encouters and interaction among the users of the building, as well as with the city surrounding it. A smart facade design provides a perfect climate through a minimum energy use and makes the building colourful and recognizable. Unfortunately the EHA and SIADR bid was not selected to be realised. Nevertheless our excellent collaboration will be continued.

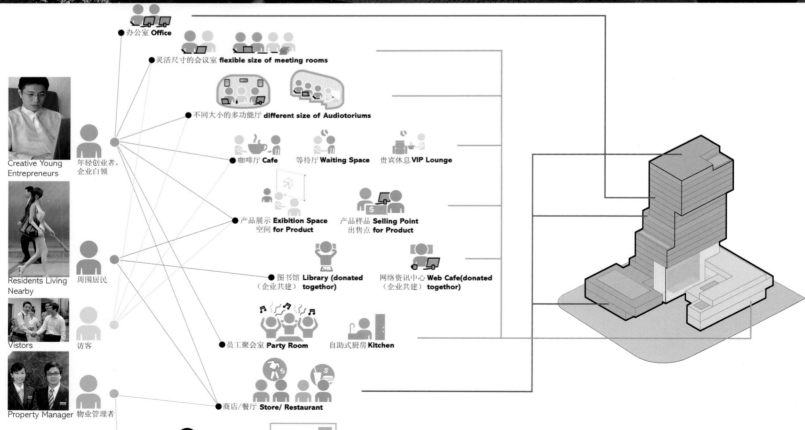

South

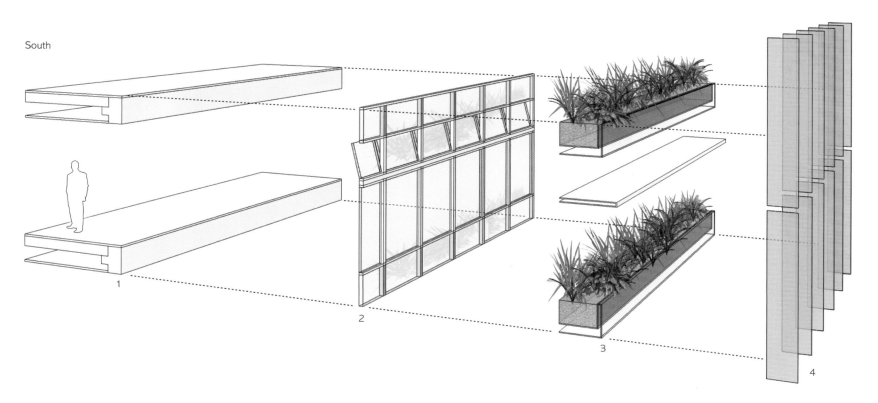

这是由荷兰事务所 Ector Hoogstad Architecten（EHA）与上海建筑设计研究院（SIADR）合作参与设计的张江孵化大楼项目竞赛方案。这个80m高的大楼构成了高新科技园区的中心，它正好处于城市中心及浦东机场之间。设计建立起不同项目元素之间强烈的联系，这些元素包含办公空间、展示空间、零售空间，从而激发了不同使用者之间的交流和互动，同时增强了建筑与周围环境的联系。一个智能的建筑表层系统提供高效的能源利用，同时使建筑美丽易识别。不幸的是，EHA 和 SIADR 竞标失败了，未能实现该方案。然而我们的优秀合作将会继续。

- New City Landmark — Mixed-use Architecture

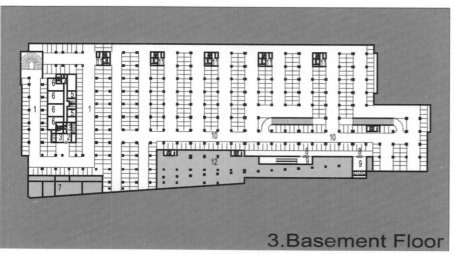

3. Basement Floor

1. HOTEL PARKING
2. ENTRANCE FROM PARKING AREA
3. HOTEL FLOOR HALL
4. SERVICE HALL
5. PERSONEL
6. STORAGE
7. WATERTANKS
8. ENTRANCE FROM PARKING AREA
9. CIRCULATION HALL OF MALL
10. PARKING AREA
11. SERVICE HALL
12. SANCTUARY

Typical Hotel

New City Landmark — Mixed-use Architecture

Architect · FEMA Architectural Co. Ltd.
Client · Gimat Birikim Inc.
Area · 80.500 m²
Program · Retail, Hotel, Convention

Ankara, Turkey

TERASMALL Outlet Center & Hotel Project

TERASMALL 批发中心与酒店

The project, which is designed in an area of 26,500 m² in the crossing of Istanbul Highway and Anadolu Avenue, consists of outlet shopping courts, catering areas, social facilities and a hotel building with 180 rooms.

Although the project was designed as a 3-storey shopping mall initially, it appeared to be unfeasible due to significant increase in the number of shopping malls in these days. Therefore, conceptual changes have been realised during the development phase of the project and certain parts of the construction area have been re-designed as a hotel. The proportions of horizontal block and vertical block have been studied diligently during the planning process and the design was completed without obstructing the construction conditions.

Shopping and the other activities are located in the horizontal block where as the vertical block comprises the hotel function. Since the higher structure is capable of a detailed visual perception of the lower block, an active and colorful composition is targeted for the top floor of the lower block.

Two different entrances to shopping areas – to basement from the southeast of the land and to ground floor from the squares formed in the south of the land – are provided by taking advantage of the slope of the land. The shopping areas are located in open streets established on these two floors. The interior circulation of the structure is provided with the various circulation elements generated at the two ends of these streets. Catering areas and social facilities are located in the "terrace" which is located on the top floor of the horizontal block enjoying the panorama of Ankara and providing the name of the project concept.

Restaurants, convention halls of various sizes and social facilities take place in the hotel together with the bedding units.

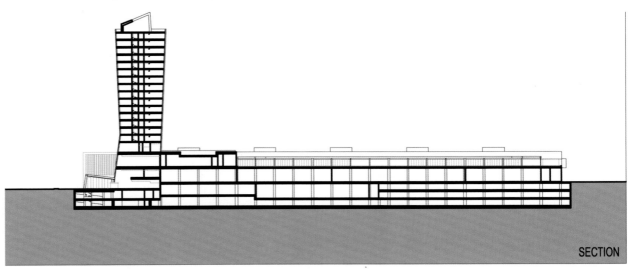

SECTION

这个项目，设计面积为 265 000 m²，位于伊斯坦布尔公路和安纳托里亚大道的交叉路口，包括批发商店大楼、餐饮区、社交设施和 180 室的酒店大楼。

尽管起初这个项目设计为一个 3 层的购物中心，但由于那段时间购物中心显著增加，这变得不可行。因此，在项目的开发阶段，开发商改变了建筑理念，施工区的一些部分重新设计为酒店。在规划阶段，建筑师对水平和垂直街区用心进行了研究，在不妨碍施工条件下，设计得以完成。

购物和其他活动区都位于水平区，酒店位于垂直区。因为在较高的建筑上能清楚的看到较矮的建筑块，较矮建筑块的顶部特别进行了充满活力的、丰富多彩的设计。

购物区有两个不同的入口，一个入口在场址西南边的地下室，另一个入口在场址南边广场的底层，它们充分利用了场址的倾斜。购物区坐落在开放的街区上，为两层。在街道两端的大量循环元素为提供了结构的内部循环。餐饮区和社交设施位于"露台"上，露台在水平区的顶楼，享有安卡拉全景，同时项目的名称被设计于此处。

不同大小的餐厅、会议厅和公共设施，通过基地单元和酒店连为一体。

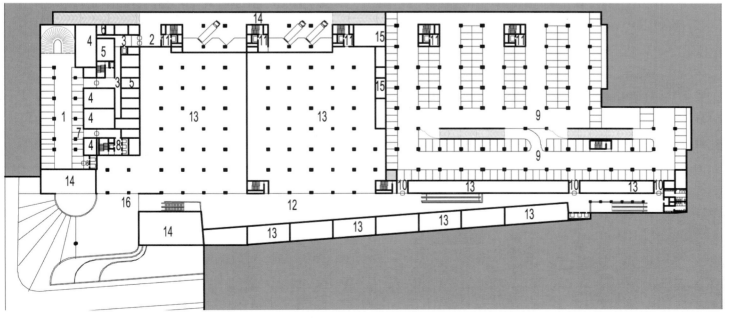

1 HOTEL PARKING
2 HOTEL SERVICE ENTRANCE
3 SERVICE HALL
4 THECNICAL AREA
5 STORAGE
6 HOUSEKEEPING
7 ENTRANCE FROM PARKING AREA
8 HOTEL FLOOR HALL
9 PARKING AREA
10 ENTRANCE FROM PARKING AREA
11 FIRE ESCAPE
12 SHOPPING STREET
13 STORE
14 UPLOADING AREA
15 THECNICAL AREA
16 LOWER ENTRANCE OF MALL

2.Basement Floor

• New City Landmark — Mixed-use Architecture

Beijing, China
South West Hotel
北京西南饭店

Architect · HENN
Client · South West Hotel
Program · Hotel, Conference, Office, Retail

The site of the five-star South West Hotel is located next to a major traffic junction of the West Fourth Ring Road and Lianhua Road in Beijing. The project is a multi-functional building complex, integrating three major programs: hotel, commerce and office. All the program units loop around a central courtyard peaking in a five-star hotel.

From the distance the overall complex creates a clearly identifiable landmark. The ascending curves of the volume are a reference to the dynamics of the context. The internal courtyard shields from the noise and the hectic city life creating a retreat for guests and customers — a shared communal habitat connecting the manifold programs.

The different spatial qualities between the exterior (city) and the interior (courtyard) are reflected in the building envelope. The inner facade appears open and transparent. Horizontal bands form balconies and outdoor gardens – secluded getaways for

New City Landmark — Mixed-use Architecture

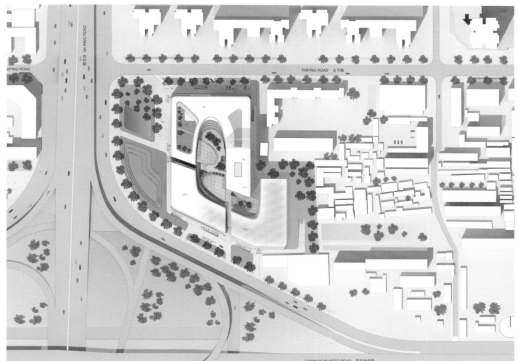

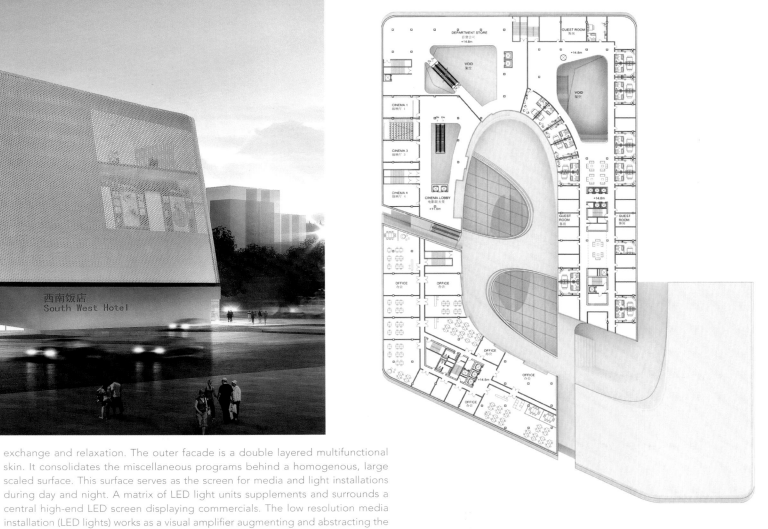

exchange and relaxation. The outer facade is a double layered multifunctional skin. It consolidates the miscellaneous programs behind a homogenous, large scaled surface. This surface serves as the screen for media and light installations during day and night. A matrix of LED light units supplements and surrounds a central high-end LED screen displaying commercials. The low resolution media installation (LED lights) works as a visual amplifier augmenting and abstracting the commercial content over the entire building facade.

Floor Plan 04

　　这座五星西南酒店的场址位于一个主要的交通枢纽，即北京西四环路与莲花路的交叉路口。这个项目是一个多功能的综合体，包括三种主要的规划：酒店、商业和办公。所有的大楼环绕着一个中心的庭院，五星级的酒店最高。

　　从远距离看，整个综合体建造了一个清晰可辨的地标。建筑体上升的曲线与动态的背景相关。内部庭院与外部的吵杂和繁忙的都市生活隔离开来，为顾客和消费者营造一个宁静的环境，作为一个连接多种规划的公共栖息地。

　　在外部城市和内部庭院的不同空间特质反映在建筑的外围上。内部立面显得开放而透明。水平带形成阳台和户外花园，有如用于闲聊和休息的"隐舍"。外部的立面是一个双层的多功能表皮。它巩固同质、大面积表层后面的各种规划项目。表层充当昼夜的媒体屏幕和灯具安装设置。一个矩阵的LED灯组合补充和环绕着一个显示广告的中心高端LED屏幕。低分辨率的媒体安装（LED灯）作为一个视觉放大器，增大和抽象化在整个建筑立面的广告内容。

• New City Landmark — Mixed-use Architecture

Qingdao, China
Qingdao Master Plan
青岛总规划

Architect · HAO / Holm Architecture Office and Archiland Beijing
Design Team · Jens Holm (HAO), Morten Holm, Tian Kun, Chen Pu, Adam Chapulski, Camilla Bundgaard, Yuxiaomin, Liulingling (Archiland)
Area · 689,000 m²
Collaborators · Archiland Beijing
Program · Movie Studios & Theaters, Office, Residential, Commercial, and Museum.

In addition to its famous Tsingtao beer, the city of Qingdao has long been a key tourist and film-production destination in northern China.

A rich mix of historic buildings makes it a sought after movie shoot location while its proximity to some of the best beaches in northern China attracts millions of tourists every year and helped its successful bid to host the Olympic Sailing competitions in 2008.

The design for the Qingdao Master Plan seeks to further develop and expand the existing elements of the city; the site is situated within the city of Qingdao and is conveniently located five minutes drive from the airport.

The site is divided into three main areas separated by existing roads. Site A is defined as a new cultural center with sites B & C comprising of mixed use residential program.

To link the three areas together, the design takes its starting point around a sunken cultural path that leads visitors through the entire programmatic experience of the new master plan. The main access to the new culture path is situated at the north

New City Landmark — Mixed-use Architecture

west corner of the site and is anchored by a new five-star hotel development. Several additional entry points to the cultural path are established throughout the site and demarcated by key landmark buildings that define the experience the nearby surroundings.

The culture path contains three courtyards creating possibilities for flexible outdoor venues, assuring a constantly changing experience for the visitor. From the individual courtyards, there is direct access to a rich blend of high-end retail, grocery stores, restaurants, movie theaters and museums.

Sites B&C create a diverse urban mix of high, medium and low-income housing set in a lush landscape. Within the residential area, community and recreational programs such as kindergartens and sports facilities are placed throughout to activate the area as a whole and create unique neighborhood experiences.

Each of the residential units in sites B&C are situated to maximize use of sunlight and natural ventilation, helping to guarantee comfortable living conditions for the future inhabitants.

"With the Qingdao Master Plan we propose a plan with a rich mix of typologies. The goal is to create a large variety of different experiences for all income groups, all benefitting directly from access to the new cultural path".

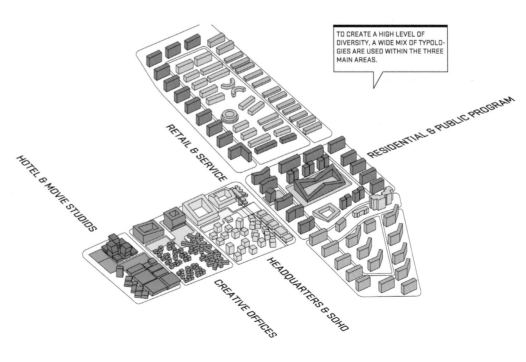

TO CREATE A HIGH LEVEL OF DIVERSITY, A WIDE MIX OF TYPOLOGIES ARE USED WITHIN THE THREE MAIN AREAS.

HOTEL & MOVIE STUDIOS
RETAIL & SERVICE
RESIDENTIAL & PUBLIC PROGRAM
CREATIVE OFFICES
HEADQUARTERS & SOHO

除了著名的青岛啤酒，青岛的城市长期以来一直是中国北方的重要的旅游地和电影制作地。

丰富的历史建筑使其成为一个广受欢迎的电影拍摄之地，而它作为中国北方几乎最好的海滩之一，吸引了数百万的游客，并帮助其成功申请举办2008年奥运会帆船比赛。

青岛总规划的设计追求进一步发展和扩大城市现有元素，场址坐落在青岛城内，到达机场非常便利，只需5分钟的车程。

场址分为三个主要区域，由现有道路隔开。场址A是一个新的文化中心，B、C场址由混合用途的住宅项目组成。

为了将三个区域连在一起，设计将一个下沉的文化路作为其起点，文化路引导游客经过这个总规划的所有主要体验场景。进入新文化路的主要路口位于场址的西南角落的，靠着一个新的五星级酒店开发体。新建了几个额外的入口点，贯穿整个场址。关键的地标性建筑将场址与附近的环境区分开来。

文化路包含三个庭院，为灵活的户外场地创造可能性，确保给游客一个不断变化的体验过程。单个的庭院里有通往一个混合功能区的直接路径，包括高端零售、杂货店、餐馆、电影院和博物馆。

B、C场址创建了一个多元化的混合住宅，在一个郁郁葱葱的景观中包括高、中、低收入三种住宅。在居民区，社区和娱乐项目如幼儿园、体育设施到处都是，将该地区打造成一个整

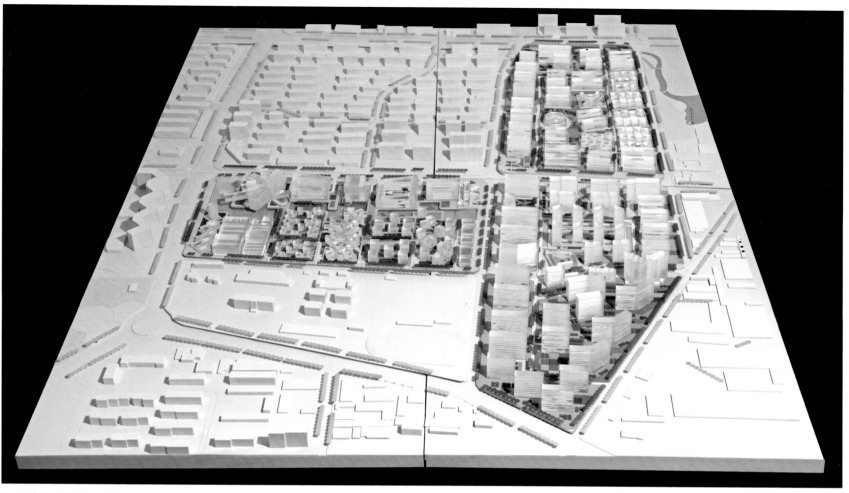

体，并创造独特的社区体验。

在B、C场址的每个住宅单元最大化利用阳光和自然通风，有助于为未来的居民保证舒适的居住条件。

"青岛总规划提出一个多种建筑种类的计划。目标是为所有阶段的收入群体建造多种不同的体验。"

New City Landmark — Mixed-use Architecture

New City Landmark — Mixed-use Architecture

Architect · KaziaLi Design Collaborative

Shanghai, China

Shanghai Cultural and Financial City

上海文化金融城

"Turning a new leaf"— A revitalization inspired by nature and history

As a previous industrial site in Southern Shanghai, Minhang district, the project focuses on creating a finance-cultural driven development. This area will revitalize and become a social beacon for the region in the future.

The project is split between office and working spaces and a unique retail environment. The working spaces include mid-rise office towers as well as high-rise office towers, the latter which serve as project keystones. The office spaces are varied and would be highly desirable for a multitude of companies and businesses. The retail environments are the most unique and attractive aspects of the project. Derived from the historical Xintiandi area in Shanghai, is the basis for one major retail attraction. Featured with European style architecture, the 3-4 storeys buildings would be ideal for local and independent cafes, bars, restaurants and product stores. Surrounding the small retail is the larger retail area, which is also derived from the highly successful Omotesando Street in Tokyo. In this area of the project patrons would walk among cutting edge contemporary architecture featuring wears from national and international fashionable brands. Just as nature follows the cycle and rhythm of seasons, so too does the site, not only organizationally but functionally. When the weather is fair nature begins to breath, so too is the site designed to expand into the exterior spaces. Dispersed amongst the raveling streets and buildings are exterior environments, landscaping moments and pocket parks.

With the projects connection and design, workers and shoppers would mingle in a natural and organic way amongst the various environments. The design concept, a leaf is something common as well as beautiful within nature. The plan design with the leaf allows for simplistic and clear organization of the site as well as a comforting environment. Further the site is a revitalization, so it is fitting the English innuendo, "turning a new leaf" or creating a fresh start. In this place a celebration of life is invoked through prosperity and what the future holds while understanding the past . As this design is a successful reflection of the site and atmosphere, we firmly believe the 125 district will be an attraction — drawing people from near and far, allowing congregation and the development of community spaces.

The Project is more than a functional and programmatic envelope. It is a place for South Shanghai that is inspired by nature and history. In this place a celebration of life is invoked through prosperity and what the future holds while understanding the past — drawing people from a near and a far.

The core is nature and a celebration of history. The form of the old buildings shape the exterior spaces allowing congregation and the development of community spaces.

"掀开崭新的一页"——自然与历史启迪下的复兴

闵行区以前是南上海的一个工业区,现在,项目旨在将其打造成一个金融—文化型街区。未来的"上海文化金融城"将崛起成为当地的一个标志性社会区域。

该项目分为办公区、工作区和一个独特的商业区。工作区包含中高层办公楼,其中高层办公楼是此项目的基石。办公空间大小不等,是诸多公司企业的绝佳选择。商业区是该项目中最具独特气息也最具吸引力的一个区域,从上海的新天地区域吸取灵感,从而作为该项目商业区的一个引人注目的亮点。欧式的建筑风格,3—4层的建筑,是当地个人经营咖啡馆、酒吧、餐馆及店铺的绝佳选择。小商业区的外部是大型的商业区,该区域的设计灵感也来源于日本东

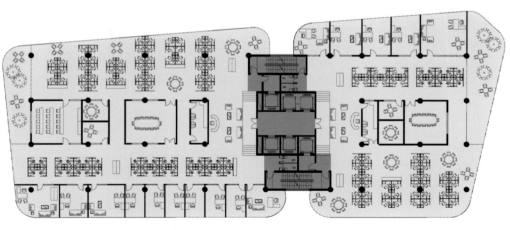

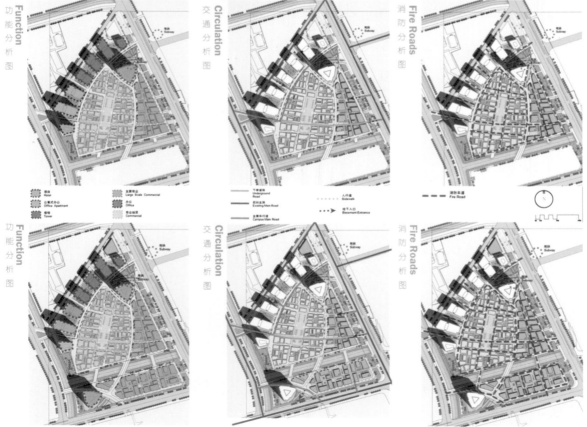

京非常成功的表参道大街。在该区域内，店主顾客们可以畅游于销售民族或国际品牌商品前卫的现代建筑中。像自然遵循四季轮回的节奏一样，本地块也如此，不仅结构上如此，功能上也是如此。气候宜人，万物复苏，设计扩展至外部空间便是寓意于此，同时，散落于充满生气的街道及楼群之间的是宜人的外部环境、景观绿化及小巧的公园。

此项目的设计，能让办公区的工作人员及购物的顾客徜徉于自然有机的、多样的环境中。设计理念来源于叶子，叶子是自然界中很平常但也非常漂亮的事物。叶子形的设计能更简单清晰地利用地块，营造一个舒适的环境，而且作为一个改造再利用项目，叶子的形式也刚好吻合了英语里的一句俗语"掀开崭新的一页"，开始新的征程。在这里，对生命的颂扬以对未来的遐想与对过去的理解而展现。设计充分反应了地块特点及区域氛围，设计师满怀信心"上海文化金融城"将会吸引来自四方的人们，发展为极具包容性与多样性的社区空间。

这不仅是一个功能型或方案性的项目，还是一个以自然与历史为灵感，为南上海而建的特别项目。在这里，对生命的颂扬以对未来的遐想与对过去的理解来表现——这些都吸引着远近人们的到来。

设计的核心思路是拥抱自然，赞颂历史。旧式建筑的形式塑造了外部空间，并为社区空间的发展提供了条件。

• New City Landmark — Mixed-use Architecture

Architect · LAVA

Berlin, Germany
THE:SQUARE³

方正

Three towers of gold, silver, and bronze and three diagonally shaped city blocks make up THE:SQUARE³, a new mixed use development inspired by sport that revitalises a unique urban quarter in Berlin. The project is just nine minutes from Alexanderplatz, the very heart of Berlin, and located near Europe's largest urban nature reserve and a sports hot spot.

Conceived by visionary developer Moritz Gruppe and designed by internationally acclaimed architects, LAVA, THE:SQUARE³ theme is Life, Nature, Sport :

1. Life: A multifunctional urban plan includes all the essentials for a high-quality and healthy urban existence for locals, workers and visitors, successfully answering the demands of a contemporary, quality lifestyle.

2. Nature: Green characterises three blocks containing apartments, retail space, a kindergarten and social services. Residents will enjoy diagonally shaped spaces, green roof — scapes with casca ding balconies, integrated garden courtyards, and overlook playing fields. Hanging plant-filled facades are articulated according to building orientation, offering an enhanced quality of living.

3. Sport: Rising above a sport "podium" are three towers of varying heights with Olympic themed metallic facades of gold, silver and bronze. Each volume is tapered to maximise sunlight, views and ventilation. Offices for sports companies and clubs, apartments, a medical and research centre, sports education facilities, a sports hotel (specifically for athletes) and a sports focused shopping mall at ground level, encircle a green piazza.

"Gone is the 'bedroom suburb', all apartments with no infrastructure. In Germany, and especially in Berlin, projects usually concentrate on one utilisation concept, for example apartments or offices. As a result there are too few supermarkets, parking spaces, jobs and school spaces. The single use concept is more common, but mixed use is the way forward," said Dirk Moritz, managing director of Moritz Gruppe.

"For us THE:SQUARE³ is more than just a development project, it's a philosophy. Living in a big city is an experience — you can't just order it online. A good mix of people, culture and lifestyles is what makes a city interesting and worth living in. Our goal is to answer the question: 'How do we want to live in the future?'"

"And we see that this blending of commercial, social, leisure and residential facilities is the solution," added Mr. Moritz.

"The sport theme runs through both the design and the utilisation concept. It references and enhances the nearby centre of excellence for high-performance sport, Sportforum Berlin, comprising Germany's largest Olympic Training Centre, by providing more facilities and services for athletes, sports officials and outdoor leisure lovers." The innovative design solution maximises the spatial experience whilst minimising the use of energy and resources. Tobias Wallisser director of LAVA said: "Re-thinking a traditional city block with different scales and typologies demanded a new approach. The building ensembles balance contemporary requirements with respect for the historic urban context.

LAVA's design integrates nature and future technologies — geometries in nature create both efficiency and beauty whilst future technologies enhance comfort."

Sustainability is embedded in the project. Mixed use ensures social sustainability. Building shapes maximise daylight, reducing the need for artificial light and energy use. Facades integrate photo-voltaics as a means of regenerative energy production. Naturally ventilated spaces minimise mechanical ventilation. Rainwater is collected and reused.

THE:SQUARE³ is one of four projects shortlisted in the prestigious 2013 MIPIM real estate awards in the Futura category.

"This recognition confirms our belief that by unifying the essentials of living — work, habitation, education, health, leisure — we will enhance the quality of life and community". Previous winners of MIPIM include the DaimlerChrysler development Potsdamer Platz (1999) and the new Reichstag (2000, special jury price).

New City Landmark — Mixed-use Architecture

一个金色的塔，一个银色的塔，一个铜色的塔，三个对角线形的城市街区共同组成了方正建筑群。体育运动让柏林城镇区活力四射，同时也启发了这个新建的多功能开发区。该项目到亚历山大广场只有9分钟的路程，临近欧洲最大的城市自然保护区和热闹的运动场所，可以说是柏林的中心。

经过视觉开发者莫丽子·格莱普构思，国际设计大师拉维的设计，方正的主题定为生活、自然、运动：

1. 生活：这个多功能的城市计划囊括了本地人、工人，游客所需高品质生活和健康城市生活必要的 元素，成功地满足了当代高品质生活方式的需求。

2. 自然：绿色是公寓、零售区、幼儿园和社会服务区这三个街区的特色。居民可以享用斜线形空间，葱郁带有大阳台并整合了花园的屋顶景致，能够俯瞰到远处的运动场。建筑的表面上挂满了植物，按照建筑的方向相连接，提高了人们的生活质量。

3. 运动：三个高度不同的塔位于一个奥运主题的运动场上方，三个塔的外表面分别为金色，

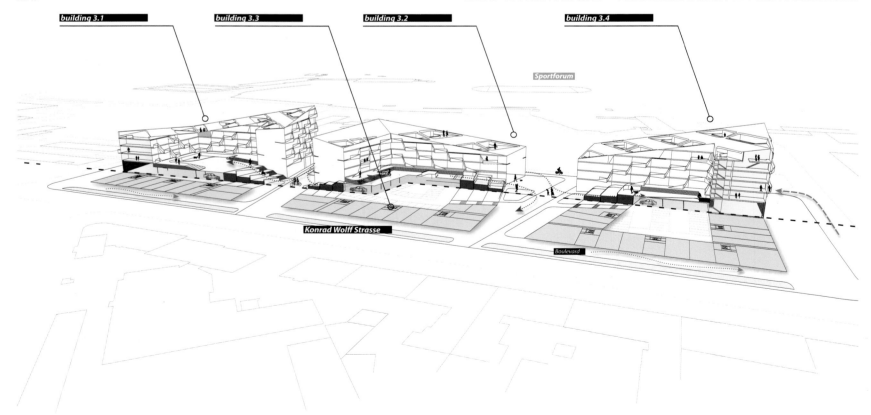

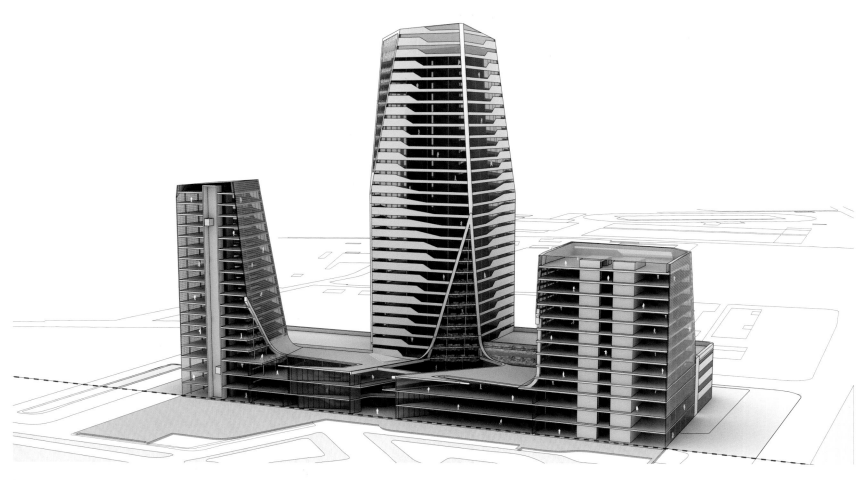

银色，铜色金属。每个塔逐渐变得尖细，最大可能地提高采光率，增大视野和通风。地面上的以体育为主的购物中心，体育公司和俱乐部办公楼，公寓，医药研究所，体育教育设施，一个体育酒店（专为运动员而设）形成了个巨大的绿色圆形广场。

"简朴指的就是宿舍，所有公寓都没有基础设施。在德国，尤其是柏林，建筑项目更多的是着重使用概念，例如公寓或办公楼。因此大型超市、停车场、工作场所、学校场所少之又少。单一功能概念在这里十分普遍，而多功能建筑稍显前卫。"莫丽子·格莱普总经理德克·莫丽子说道。

"对于我们来说，方正不仅仅是个开发区项目工程，它更是种哲学。生活在大都市是种经历——你不能在网上订购这种经历。民族、文化、生活方式在这里交融，城市因此而趣味十足，值得我们住在这里。我们的目标是解答这个问题'：将来的我们想要什么样的生活？'"

"我们找到了答案即商业、社会、休闲、居住合为一体的设施。"莫丽子先生又说道。

"该项目的设计和使用概念都贯彻体现运动这一主题。附近高级体育中心即柏林体育馆为

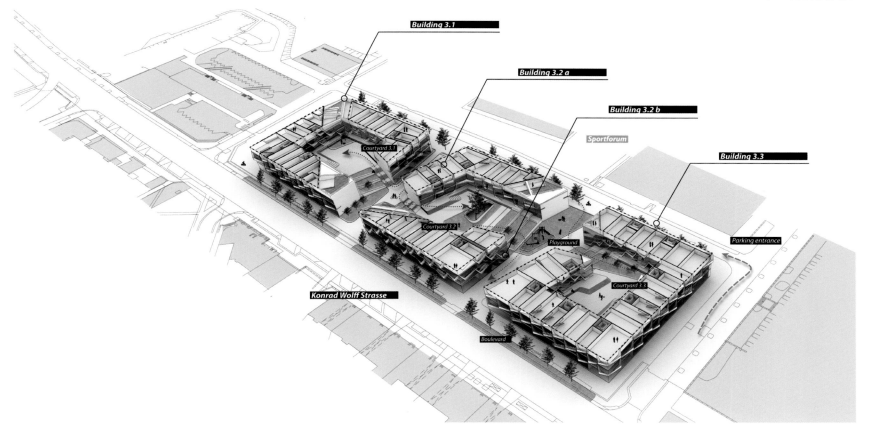

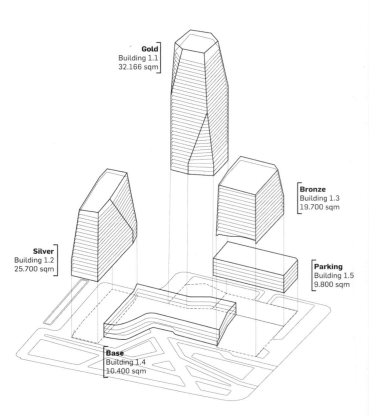

Gold
Building 1.1
32.166 sqm

Silver
Building 1.2
25.700 sqm

Bronze
Building 1.3
19.700 sqm

Parking
Building 1.5
9.800 sqm

Base
Building 1.4
10.400 sqm

Plot 1
Plot Size 35.900 sqm
Footprint 9.600 sqm
Total Floors Area 100.000 sqm

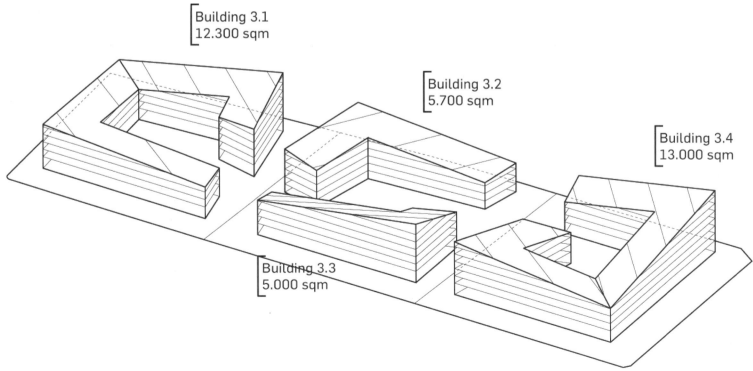

Building 3.1
12.300 sqm

Building 3.2
5.700 sqm

Building 3.4
13.000 sqm

Building 3.3
5.000 sqm

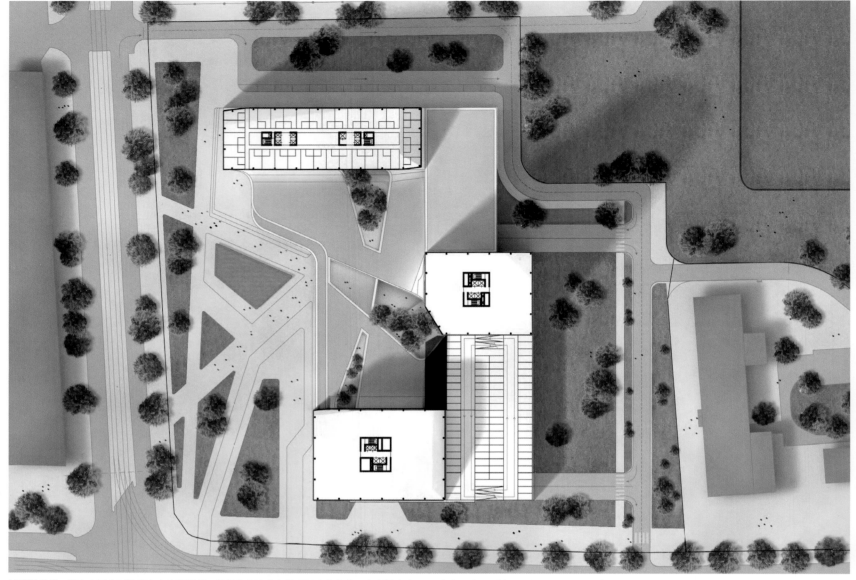

其提供参考,反之方正也对体育中心有所提升,柏林体育馆内有德国最大的奥林匹克训练中心,为运动员,体育裁判,户外运动爱好者提供更多的设施和服务。"这种创新式的设计解决方案在将空间体验最大化的同时将对能源和资源的使用降至最低。拉维公司的总管托拜厄斯·沃利斯说道:"重新考虑不同规模和象征的传统城市街区需要一个新的方式。该建筑巧妙地平衡了当代需求和对城市历史环境的尊重。"

拉维的设计将自然融入了未来科技——大自然的几何学带给我们的是效率和美丽,而未来科技则提升了我们的舒适度。

可持续发展深深得印于这项工程,多功能保证了社会的可持续性。建筑的形状将采光率最

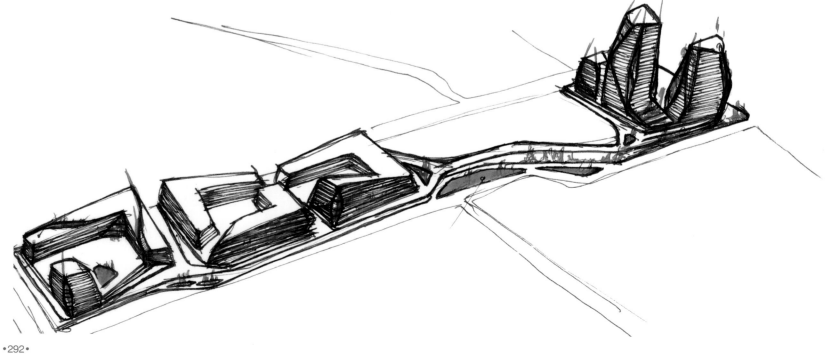

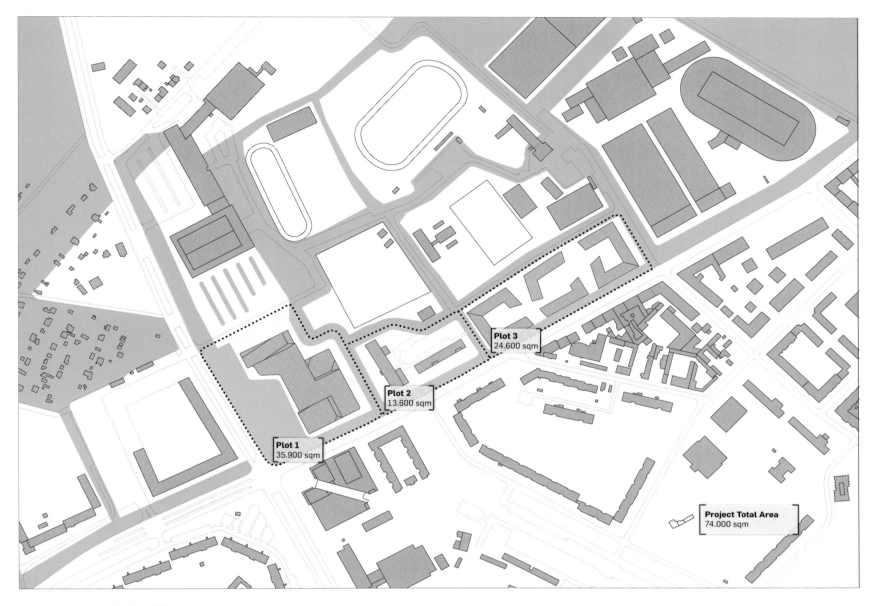

大化,减少了人工灯光和能源的使用。建筑表面运用了光电池作为能源再生的方式。自然的通风尽可能地减少了机械通风的使用。雨水则被收集起来再次利用。

方正是2013年MIPIM房地产奖未来这一方面的4个候选项目之一。

"通过统一生活必需品——工作、居住、教育、健康、休闲——我们将会提高生活和社区质量,对方正的认可无疑正视了我们这一概念。MIPIM奖历届获得者包括波兹坦·普拉兹的戴姆勒克莱斯勒开发区(1999)和新德国国会大厦(2000年,评委会特别价格)。

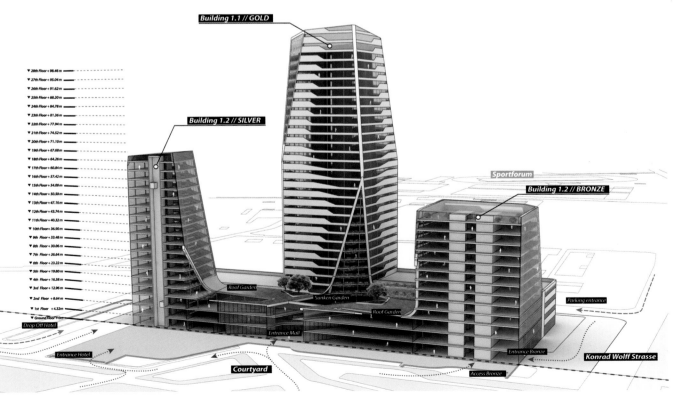

- New City Landmark — Mixed-use Architecture

Architect · Perkins Eastman
Client · Flushing Commons, LLC

Queens, New York, USA

Flushing Commons

法拉盛综合体

A need for public open space provided the impetus for this master plan for a 167,225 m² mixed-use development to revitalize a municipal parking lot in the heart of downtown Flushing in Queens, NY. The program includes 570 condominiums, retail, commercial space, a community facility, YMCA, and a multi-level underground parking garage for 1,600 vehicles (replacing the existing surface-level parking while accomodating new uses). The new 6,070m² town square provides public gathering and event spaces and enhances pedestrian circulation with access from nearby mass transit.

The vision for the development includes four buildings organized around a central elliptical green comprising lawn, plazas, gardens, and water features. The massing of the buildings minimizes shadow conditions on key paths and open spaces, reinforcing the creation of an urban destination and place to congregate.

As a reflection of the existing land use patterns adjacent to the site, the development complements the existing street-facing conditions through a thoughtful mix of street-level retail and commercial development. As a benchmark of design excellence in urban design and sustainable architecture, the development will pursue a LEED® Silver Certification.

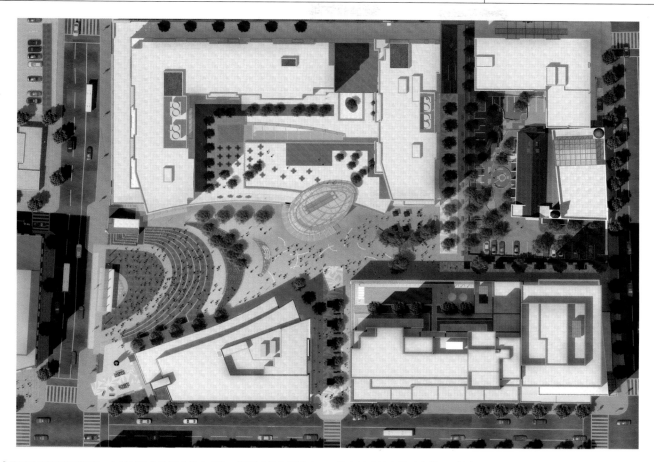

一个 167 225m² 的综合开发项目需要一个公共开放空间来壮大声势，以振兴位于纽约皇后区法拉盛市中心的一个市停车场。项目规划包括：570 室的公寓、零售、商业空间、社区设施、基督教青年会和供应 1 600 个车位的多层地下车库（替换现有的表层停车场，同时提供新的用途）。6 070m² 的新城市广场提供公共集会和活动空间，增加与附近公共交通联系的人行道。

该综合体包括四栋大楼，它们环绕着一个中心椭圆绿色草坪、广场、花园、水景。大片集中的建筑在重要大道和开放空间上形成的阴影区，试图将该综合体建造成人口集聚的都市目的地。

为了体现场址附近现有土地的利用模式，开发商经过深思熟虑建造临街零售和商业建筑，以完善街立面的现有状态。作为卓越的城市设计与可持续建筑的楷模，该综合项目将追求 LEED·银级认证。

• New City Landmark — Mixed-use Architecture

The Wharf Southwest Waterfront

西南滨水码头

Architect · Perkins Eastman DC / EE&K a Perkins Eastman Company, Rockwell Architecture Planning & Design, PC
Client · Hoffman-Madison Waterfront
Location · Washington, DC, USA
Program · Residential, Office, Retail, Hotel, Cultural

The Wharf is a new mixed-use waterfront neighborhood being developed by Hoffman-Madison Waterfront as part of the District of Columbia's Anacostia Waterfront Initiative. The site is located on the historic Washington Channel, situated along DC's southwest waterfront and adjacent to the National Mall. The waterfront redevelopment stretches across 109,265m² of land and 97,125m² of water, from the Municipal Fish Market to Fort McNair.

When complete, The Wharf will comprise approximately 297,290m² sf of new residential, office, retail, hotel, cultural, and public uses including parks, cultural centers, promenades, waterfront piers, and docks. It will be a vibrant mixed-use community comprising of a series of "places", varying in scale, program and design. Together, they create a waterfront destination for both local residents and visitors alike.

The architectural character relies on a diversity of scales and materials, utilizing stepped-back facades, a variety of complementary materials, and careful attention to the pedestrian scale. The Wharf responds differently to the street and waterfront elevations, offering significant views to the water while responding to the existing architectural scale and character of the District. The project will be among the first in the District of Columbia to be LEED® ND-Gold and will include a very innovative stormwater management system and electrical cogeneration plant, while all buildings will strive for LEED Silver certification.

New City Landmark — Mixed-use Architecture

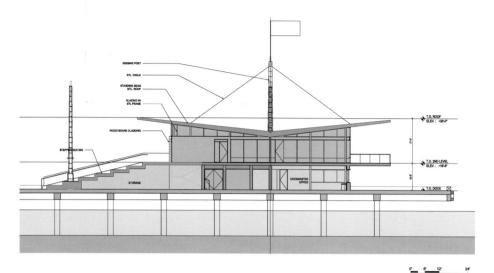

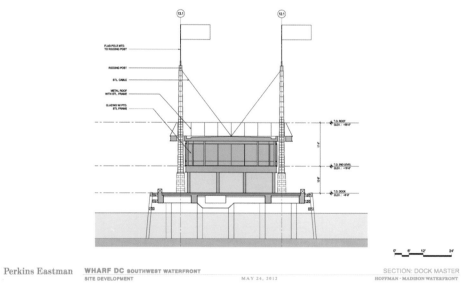

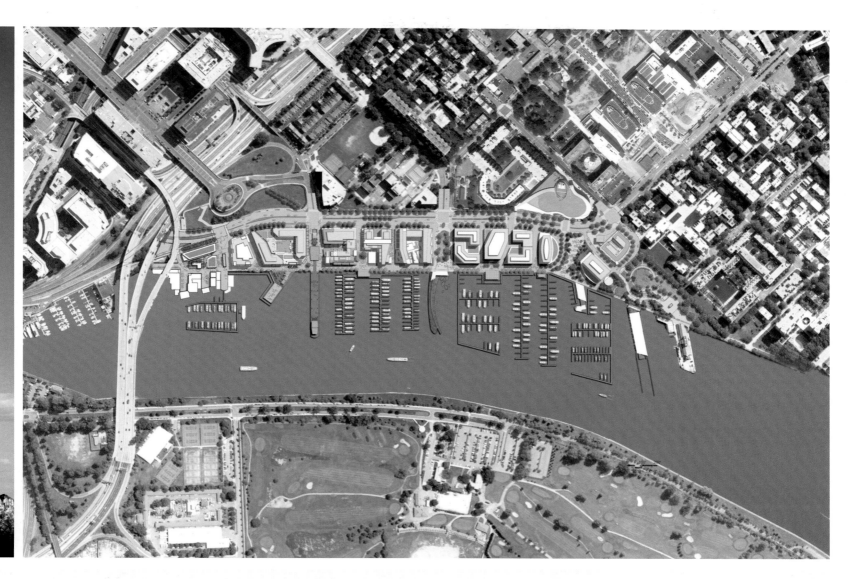

　　码头是一个混合用途的滨水发展项目,由霍夫曼麦迪逊滨水区集团作为哥伦比亚特区 Anacostia 滨水区倡议的一部分负责。场址位于历史悠久的华盛顿海峡,坐落在哥伦比亚特区的西南海滨沿线区域,毗邻国家广场(National Mall)。开发区域从鱼市场到麦克奎尔堡,横跨 109 265m² 土地和 97 125m² 水域。

　　完成后,码头将有大约 297 290m² 的新住宅、办公室、零售、酒店、文化和公共空间,包括海滨公园、文化中心、海滨长廊、防洪堤和港区。这将是一个充满活力的综合社区,由一系列不同规模、规划和设计的"地方"组成。他们一起为当地居民和游客创建了一个海滨之地。

　　建筑风格依赖多样性的规模和材料,充分利用后退外立面、各种各样的补充材料和对行人规模的仔细关注。码头分别对待不同高度的街道和滨水区,为水面提供重要的景观,同时关注现有建筑规模和区域特征。该项目将成为哥伦比亚特区首批获得 LEED 黄金认证的建筑,将包括一个非常创新的雨水管理系统和电热电厂,而所有建筑物将争取 LEED 银级认证。

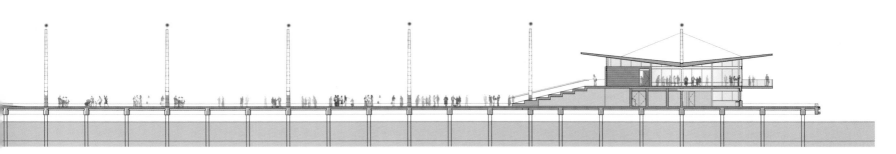

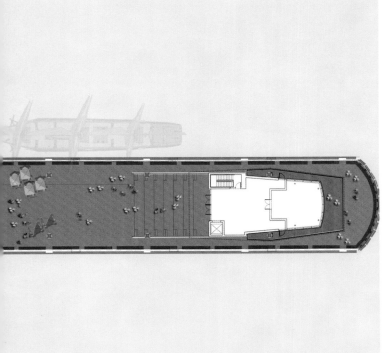

Perkins Eastman — WHARF DC SOUTHWEST WATERFRONT
SITE DEVELOPMENT — JULY 12, 2012

DOCK-MASTER
HOFFMAN - MADISON WATERFRONT

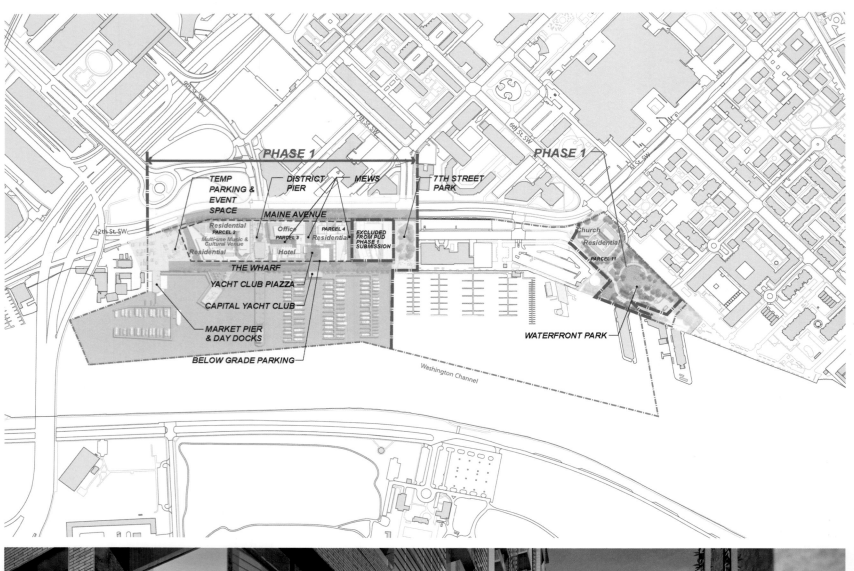

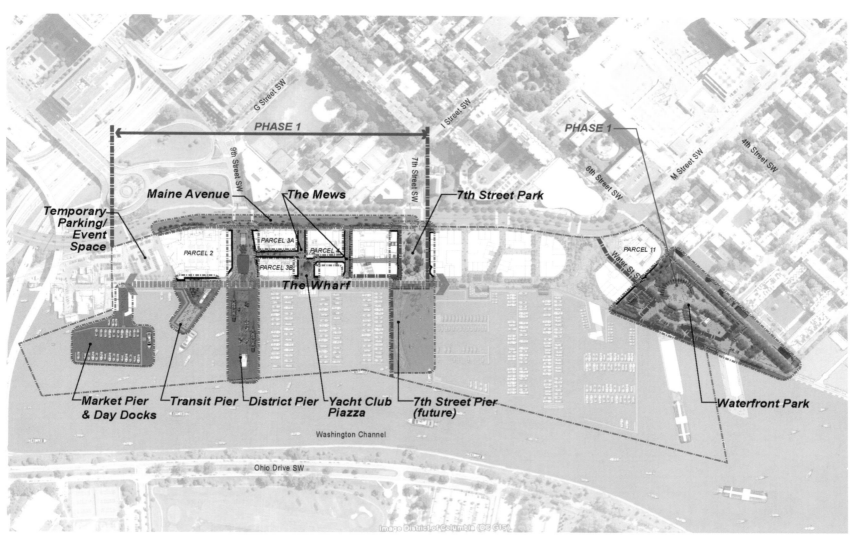

• New City Landmark — Mixed-use Architecture

Architect · Progetto CMR
Client · Municipality of Tianjin
Area · 155,000 m²
Program · Retail, Residential, Entertainment

Tianjin, China

Tianjin Jie Fang South Road Commercial Center

天津解放南路商业中心

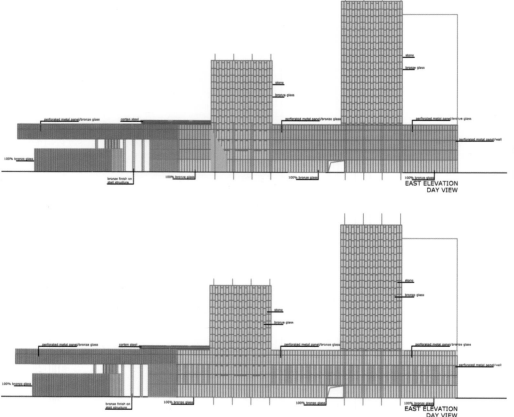

The project, located in Jie Fang South Road, in the southern part of Tianjin, consists of a multifunctional complex which combines retail, commercial, entertainment and residential activities.

The design is conceived to achieve an effective integration between human and nature. The stylistic choices of the four towers and the podium convey an immediate feeling of modernity, thanks to the dynamic twist of glass, metal, colors and stone, and to the peculiar

cladding chosen for the facades.

Another key feature of the whole design process is the emphasis put on the human dimension. At the very entrance of the podium, which hosts all the commercial activities, a covered square welcomes the visitors, representing an ideal conjunction of the project axes. The entertainment building and the three service-apartment towers embrace an internal garden, a courtyard specifically designed to improve circulation and to provide the visitor with an intimate space for their interactions.

Crucial attention has been dedicated to sustainability. The green roof, covering the podium, is not only a place for social life but also and more importantly a solution to improve the microclimate and environmental well-being. Moreover, materials, technologies, and building orientation have all been carefully analyzed in order to minimize the impact on the surrounding environment.

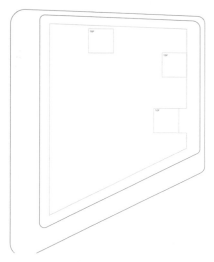
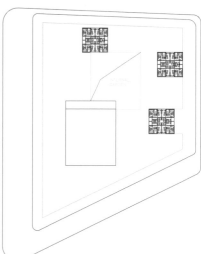
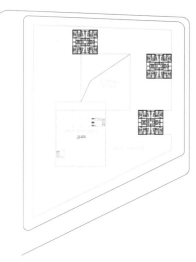

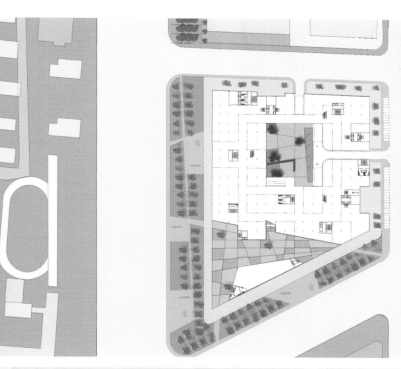
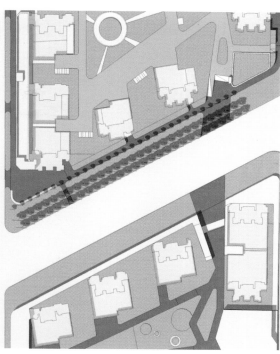

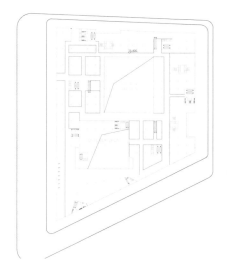
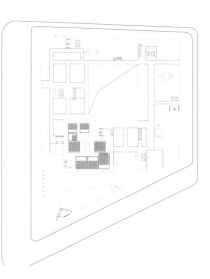
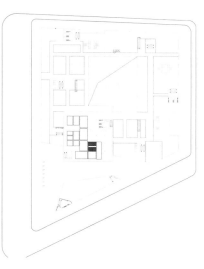

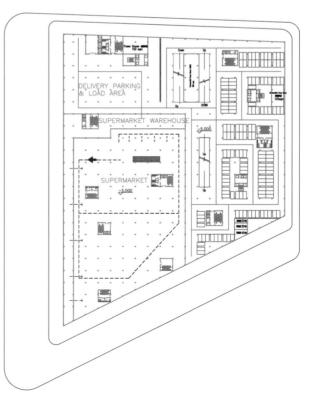

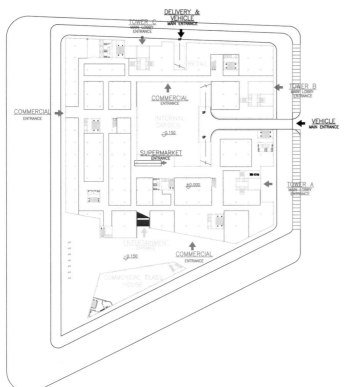
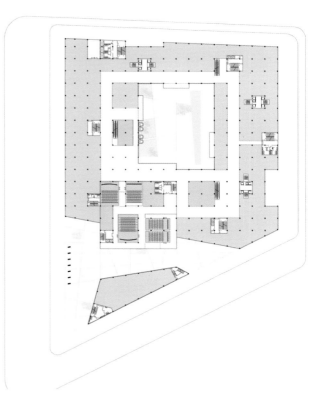

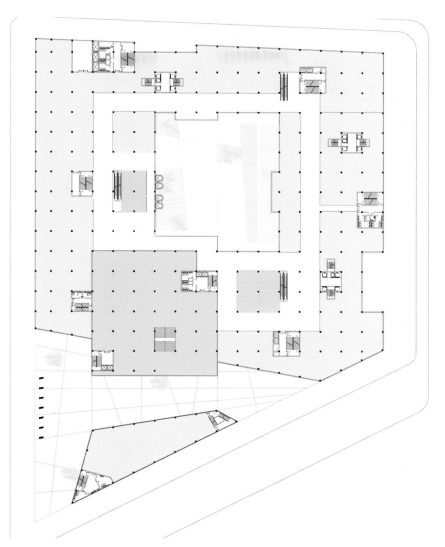

该项目位于天津南部的解放南路，是一个多功能的综合体，包括零售、商业、娱乐设施和住宅。

设计实现人与自然的有效融合。四座大楼和裙楼的风格选择如动态蜻蜓的玻璃、金属、色彩、石材以及特别的外立面幕墙，使得综合体传递出一种明显的现代感。

整个设计的另一个重要特征是对人文情怀的强调。在举行所有商业活动的裙楼的入口，一个绿化广场迎接游客，这里是综合体轴线的理想交界处。娱乐大厦和三座酒店公寓大楼环绕着一个内部花园，量身定制的庭院改善循环路线，并为游客提供一个用于交流的私密空间。

建筑师特别关注建筑的可持续持性。裙楼顶部的绿化花园不仅能作为社交区，更是改善小气候和环境的一个良好方案。此外，建筑师仔细地分析材料、技术、建筑方位，以减少对周围环境的影响。

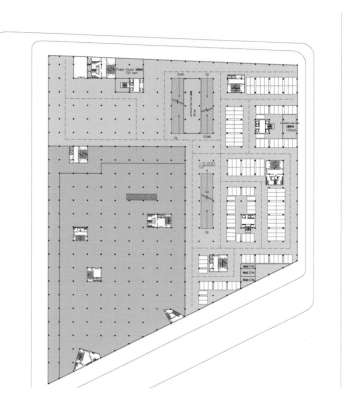

- New City Landmark — Mixed-use Architecture

Chengdu, China
9 Square Shopping Center
九方购物中心

Architect · RTKL Associates Inc.
Client · Shenzhen CATIC Real Estate Development Co., Ltd.
Area · 78,000 m²
Photography · RTKL/David Whitcomb

Taking advantage of its location within the civic district and adjacent to a city park, 9 Square Shopping Center in Chengdu presents two different facades — a bolder, streamlined elevation facing the primary roadway and a layered, varied elevation opening onto outdoor plazas connecting to the park. While the design of the building is contemporary and efficient, the team also wanted to reflect the city's rich heritage through this 78,000 m² retail, entertainment and office project. Special exterior perforated metal panels were created to celebrate Chengdu's famous woven brocades. The panels incorporate patterns found in these traditional weavings and create a subtle texture on the façade during the day; at night, they are back-lit with changing color to evoke a dramatic, sophisticated atmosphere.

This reference to the Chengdu brocade is continued and explored throughout the interiors as well. The layout is comprised of an elliptical center court with two legs on either side, while the ceiling of the curved side of the mall is highlighted by a sweeping band of patterned punched and back-lit metal, recalling similar details to the building's facade. Countering the glass handrail, the straight side of the mall is laminated with colored bands of translucent silk fabric that flow up through the space, connecting the various floors and highlighting the form and materials of the brocade. In the center court, crisscrossing lines of light and fiber optics run vertically along the bulkhead and create a loose web of woven threads to speak to the structure and the craft of making the brocades. Additional detail of pattern and form from the underside of the escalators to the vanities in the restroom speak to the flowing form and texture of brocade fabric.

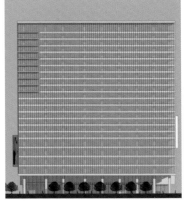

坐拥地处公民生活区内，毗邻城市公园的绝对地理优势，九方购物中心展现了两种不同的外观——一种是大胆、流线型的外表俯瞰主干道；一种是多层的、多样化的外观面对连接到公园的户外广场。虽然建筑的设计是现代化的，是高效的，但是设计团队还是想通过其78 000m²的零售，娱乐和办公项目折射出该城市丰富的文化遗产。特殊的户外用品金属板已经制作出来以庆祝成都著名的编织锦缎。面板上有这些传统的编制图案白天在外墙上显示出微妙的纹理，到了晚上，背光面将会改变颜色，营造一种精致、戏剧性的氛围。

这种与成都锦缎相关联的做法也同样延续并开发到室内。该布局由在两边都有两条腿的椭圆形中庭构成，同时，在购物中心弧形边的天花板一侧，被花纹穿孔的扫帚图案以及背光金属片所突出，让人联想起建筑外围的类似细节。与玻璃扶手截然不同的是，商场的直线边层压着丝绸织物做成的彩带，飘逸于空间里。连接繁多的楼层，突出锦缎的材质和形式。在中庭，纵横交错的光线和纤维反射光沿着隔板纵向乱串，形成松散的编织网络与结构和锦缎工艺交相辉映。至于其他额外的形式和模式细节就是，从电梯底部到洗手间的洗脸台，一路上都在述说着锦缎的飘逸与质地。

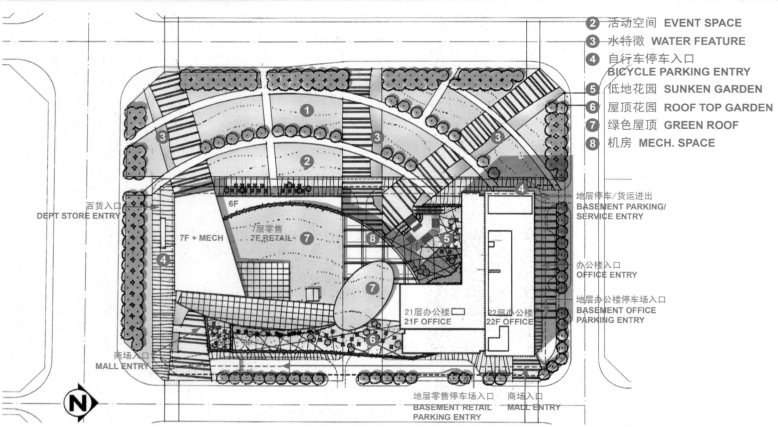

❷ 活动空间	EVENT SPACE
❸ 水特徵	WATER FEATURE
❹ 自行车停车入口	BICYCLE PARKING ENTRY
❺ 低地花园	SUNKEN GARDEN
❻ 屋顶花园	ROOF TOP GARDEN
❼ 绿色屋顶	GREEN ROOF
❽ 机房	MECH. SPACE

百货入口 DEPT STORE ENTRY
地层停车/货运进出 BASEMENT PARKING/SERVICE ENTRY
办公楼入口 OFFICE ENTRY
地层办公楼停车场入口 BASEMENT OFFICE PARKING ENTRY
商场入口 MALL ENTRY
地层零售停车场入口 BASEMENT RETAIL PARKING ENTRY
商场入口 MALL ENTRY

7F + MECH
6F
7层零售 7F RETAIL
21层办公楼 21F OFFICE
22层办公楼 22F OFFICE

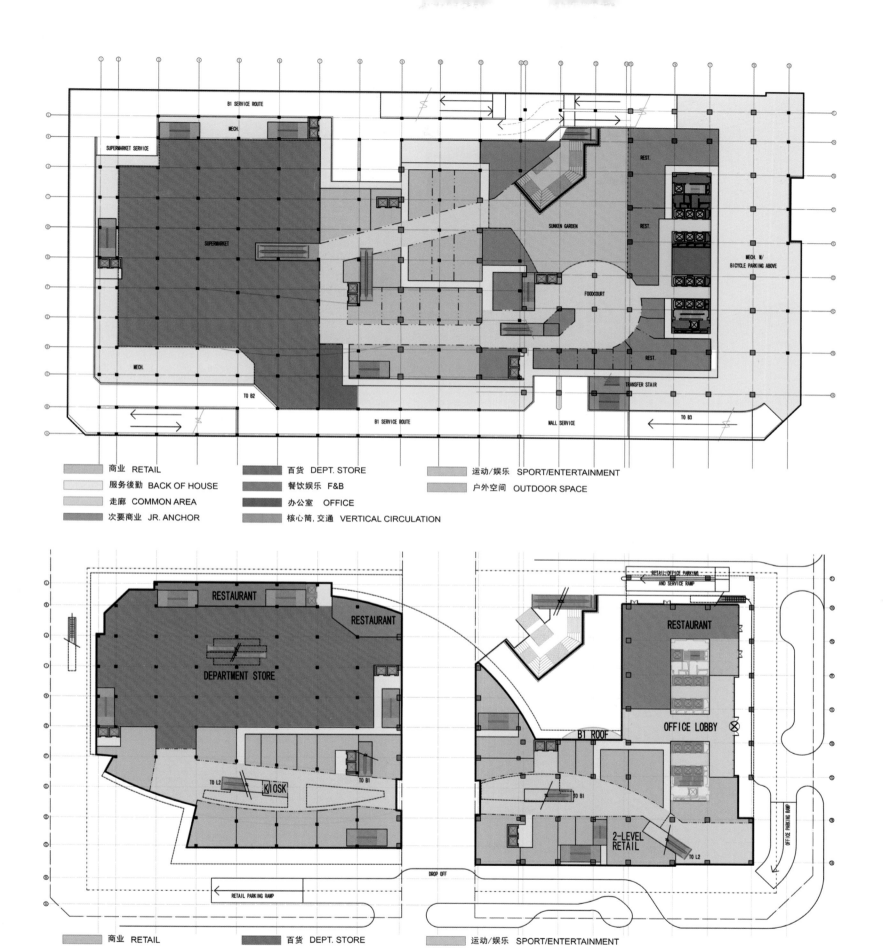

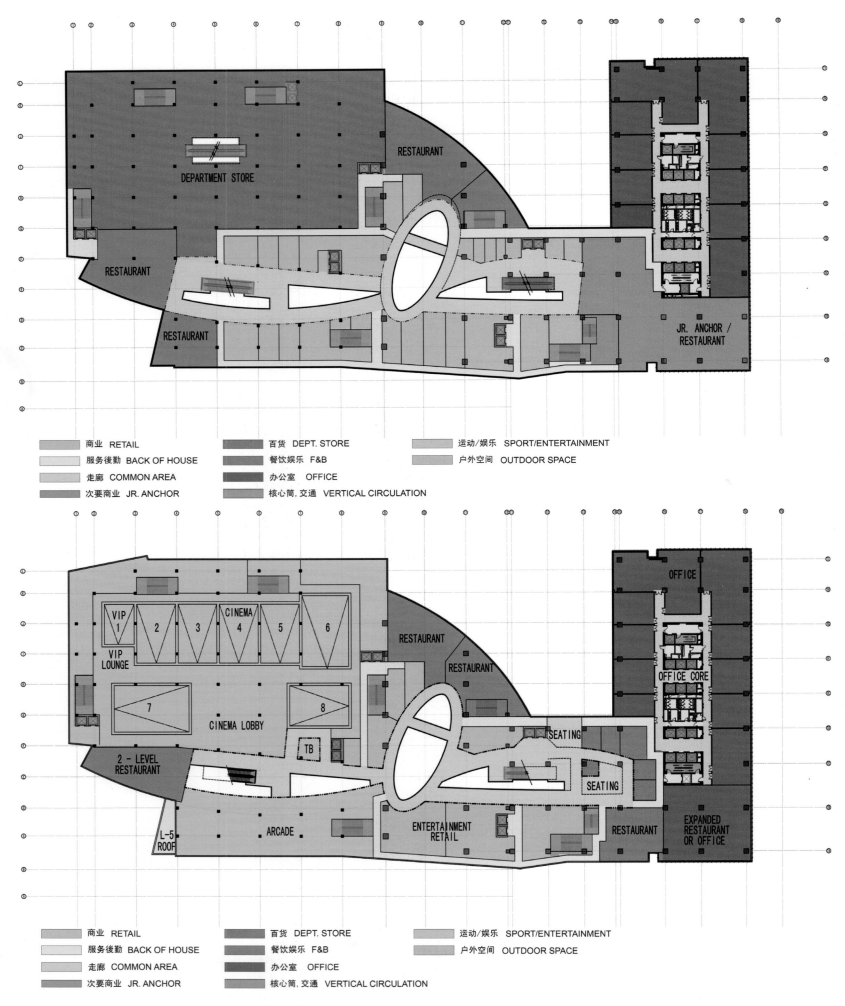

• New City Landmark — Mixed-use Architecture

Architect · SPARK
Client · CapitaLand China Holdings Pte Ltd
Local Design Institute: Sunlight Architects and Engineers
Project Director: Jan Felix Clostermann
Area: 150,000 m^2

Photography: Fernando Guerra, Christian Richters, Shuhe

Ningbo, China
Raffles City Ningbo

宁波来福士广场

Construction is underway on SPARK's landmark Raffles City Ningbo development, a one-stop shopping, dining, business, and lifestyle destination covering some 157,800 m^2.

In order to realize the project, a world-class design team is assembled, led by award-winning architects SPARK, and including Meinhardt (MEP consultant), Arup (fire engineer), MVA (traffic consultant) and the Ningbo local design institute (LDI).

Ningbo is home to China's second largest port, boasting an increasingly affluent population. Raffles City Ningbo was launched last year, graced by Singapore Prime Minister Mr. Lee Hsien Loong and Zhejiang Province Vice-Governor Mr. Gong Zheng. The project, which comprises a mall, a Grade A office tower and serviced residences, is the latest in the "Raffles City" brand developed by Singapore-based CapitaLand.

Raffles City Ningbo is located to the east of the Yuyao River, within the Jiangbei district, next to Ningbo's historical downtown. The project marks the final phase of CapitaLand's masterplan and will act as a catalyst development for the Jiangbei District, creating a new downtown district in the city. The planned subway link beneath the office tower will cement the development into the city's infrastructure and assist in the creation of this new shopping and business destination.

According to Stephen Pimbley, CEO of SPARK:

"In order to meet the needs of the full spectrum of end users, the development has a 'slow domestic face' and a 'fast civic face'. The residential component sits on the corner of a landscaped, sunken courtyard: this helps moderate and articulate the scale of the tower and creates a calm domestic environment, sheltered from the noise of the city by the mass of the retail podium." He continues: "The animated face of the retail podium and the office building engage directly with the city and link the older part of Ningbo across the river with the developing quarter, anchored by Raffles City."

Although Raffles City Ningbo is a large-scale development by any standards, SPARK has designed the various components to work on a human scale. Wavy, layered ribbons, reminiscent of the strata found in rock formations are used throughout.

Pimbley explains: "This is not just an architectural device to break up the mass of the building facing the residential tower; it has become something of a letmotif, tying all aspects of the development together. In the context of the retail podium, for example, this applied, ribbon aesthetic enables each part of the structure to fit seamlessly together, lending the building a feeling of an almost domestic scale, so often lacking in the ubiquitous shopping mall box."

The retail podium has already been awarded a Green Mark Gold Badge by Singapore's Building and Construction Authority. Pimbley says, "The building balances spatial demands with sustainable requirements. Excessively high spaces, beloved by many mall developers, are perceived to add quality and value. In my view, they create extra building volume which requires more energy to heat and cool and therefore can be wasteful and unsustainable."

正在施工中的宁波来福士广场是一个集购物、餐饮、商业、休闲于一体的综合体项目，约157 800m²。

为了完成这个项目，开发商组建了一个由SPARK领导的世界级设计团队，包括Meinhardt、Arup 、MVA和LDI。

宁波是中国的第二大港口，拥有越来越多的人口。宁波来福士广场于去年开始兴建，新加坡总理李显龙先生和浙江省副省长龚正先生对该项目表达了致意。该项目是新加坡嘉德置地开发的"来福士广场"品牌中最新的，包括一个购物中心、一个甲级写字楼和服务公寓。

宁波来福士广场位于江北地区的余姚河的东岸，紧邻宁波市区的历史中心。该项目标志

着嘉德置地总体规划的最后阶段，将作为江北地区开发的一个催化剂，促进一个新商业区的创建。办公大楼下规划中的地铁路线将巩固城市的基础设施，并有助于创建新的购物与商业目的地。

据SPARK的首席执行官Stephen Pimbley所说："为了满足终端客户全方位的需求，项目有'慢节奏的居家生活'和'快节奏的都市生活'两种环境。住宅坐落在一个下层景观庭院的角落：这有助于清楚的呈现住宅大楼，营造一种平静的居家环境，与嘈杂的大片零售商场区隔离开来。"他继续说："生机勃勃的零售商店和办公建筑与城市融为一体，并跨河将宁波较为古老的部分与新开发区连为一体。"

虽然以任何标准来看宁波来福士广场都是一个大规模的综合体，SPARK 仍然设计了各种很人性化的组成部分。让人想起岩层中地层的波浪的分层丝带贯穿始终。Pimbley 解释说："这不仅仅是一个建筑设备，驱散群众面向住宅大楼的巨大感，也将建筑体的所有方面的连接在一起。在零售商店的背景中，例如，运用的丝带审美使每个部分的结构无缝对接，给大楼注入一种家庭氛围感，让你常常沉迷在无处不在的购物中心盒子里。

零售商店获得了一个新加坡市建设局颁发的绿色标记的金徽奖。Pimbley 说："建筑平衡了空间需要与可持续需求的关系。许多购物中心开发商喜欢的特别高的空间被认为是给建筑体增添品质和价值。在我看来，他们创建额外的建筑体积，需要更多的能源来供热和冷却，因此是浪费的、不可持续的。"

- New City Landmark — Mixed-use Architecture

Architect · The Jerde Partnership, Inc
Client · ENKA TC
Area · 235,000 m²
Program · Retail, Entertainment, Residential, Office, Cultural

Moscow, Russia
Kuntsevo Plaza

昆士伏广场

Situated within central Moscow, Russia, the new integrated mixed-use Kuntsevo Plaza will deliver a modern community gathering destination rooted in art, nature, and urban connectivity. Providing a new stage for dynamic public activity and distinct commercial offerings, the pedestrian-oriented center will establish a vibrant leisure, shopping, business, and residential complex reconnecting the urban fabric of the historic Kuntsevo district, while creating a new landmark for the city. The project's key urban design principles establish vital connections to the surroundings, including to the nearby transit line with a grand public plaza at the front entry closest to the metro, and multiple entryways and accessibility from the various streets, taking advantage of the site's unique topography and grade change. Consisting of approximately 235,000m² of GFA, including 70,000m² of spectacular light-filled retail and entertainment spaces, plus three contemporary designed Class-A office buildings, and two high-rise residential towers with lush rooftop park terraces, the new complex will deliver an innovative and one of the first truly integrated mixed-use destinations in Moscow that will attract regional visitors and serve as a catalyst for future sustainable development.

The project's design enhances its potential to become a continuously active public realm. Through its synergistic combination of retail, entertainment, dining, cultural uses, state-of-the-art office and modern residential, within a single, world-class facility, all designed at an experientially articulated pedestrian scale, Kuntsevo Plaza will further enhance its respected, historic city, breathing new life into Moscow upon its expected completion in April 2014.

New City Landmark — Mixed-use Architecture

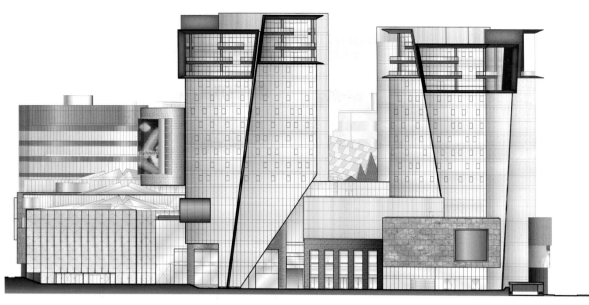

Evolution

坐落于俄罗斯莫斯科市中心，新的综合性的昆士伏广场将提供一个根植于艺术，自然与城市融合为一体的现代化聚集地。给充满活力的公共活动和独特的商业产品提供了一个新的平台，行人导向中心将集合历史悠久的昆士伏建筑风格，集充满活力的休闲、购物、商务、住宅于一体，成为城市新地标。该项目的关键城市设计原则是与周边环境建立重要联系，包括附近的运输路线，紧靠地铁的大型公共广场前门以及多个出入口，并且要利用该场地的独特地形和品味变化与周边各类型街道无障碍连接。约235 000m² 的建筑面积，包括 70 000m² 的壮观且光线充足的零售和娱乐空间，加上三个当代设计的 A 级写字楼，两幢高层住宅大楼与含梯田的郁郁葱葱的屋顶公园，新的综合广场将传达一种创新意识，成为在莫斯科的第一个真正意义上的综合广场，将会吸引更多地区游客，成为可持续发展的催化剂。

该项目的设计增强了它的潜力，使它成为一个持续活跃的公共场所，通过其协同零售、娱乐、餐饮、文化用途以及最先进的办公室和现代住宅，成为拥有独立的、世界一流的设施。所有设计都按照人流量严格执行，昆士伏广场将进一步提升受人尊重的历史名城的知名度，将在 2014 年 4 月竣工之后给莫斯科成注入新的活力。

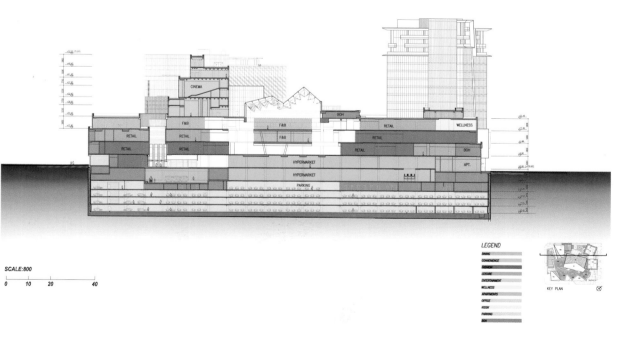

• New City Landmark — Mixed-use Architecture

Architects · J2-DESIGN
Designer · Feng Houhua, Xie Guanrong
Area · 1,100,000 m²

Guangzhou, China

Panyu Wanbo CBD Commercial Plaza

番禺万博 CBD 商业广场

Panyu Wanbo CBD commercial plaza project is located at the heart of Panyu CBD core region. This project, as the largest urban synthetical block in Panyu, also is a key program of Wanda Group in south China, in 2003.

Compared with Wanda squre, Panyu CBD commercial plaza represents the strategy of Wanda urban synthetical block's upgrading:

1. The re-scheme of spatial arrangement. Compared with Wanda square's traditional spatial structure — two courts and one street, separately, interior space of Panyu Wanbo CBD commercial plaza is naturally integrate. The connection between court and straight street, and the distinguished shape of the bridge make interior space more unique.

2. It's a full display of the promotion of the demand and style of Wanda plaza theme. The site has cradled Lingnan culture. As a consequence, light brown and black curved roofline becomes a primary manner to show the theme. The black iron lines defined by slight brown interweaves a glamour area featuring the modern Panyu Wanbo and the luxury and tranquility of local site.

3. the enhancement in the design standard of humanity and security. Panyu Wanbo CBD commercial plaza is a high-standard project in Panyu Wanda CBD commercial area. Huge improvement in daily facilities, in addition to the guarantee of fire safety of structure and sub-district, will need the materials being: Antibacterial texture aluminum, GRG plates, metal panels and other decorative materials, which all can keep luxury and security. This is constantly the model of the high-standard project of Wanda.

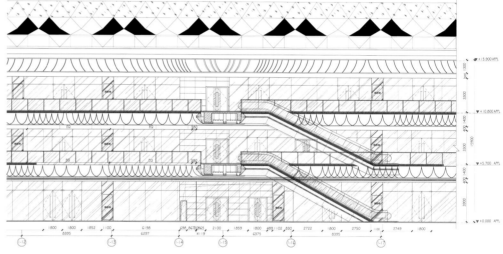
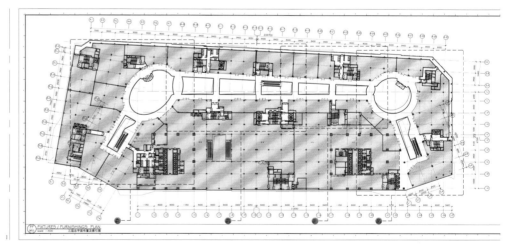

　　番禺万博 CBD 商业广场规划选址位于番禺 CBD 核心地域，作为番禺 CBD 区的最大的城市综合体，该项目亦是万达集团 2013 年在中国华南区的重点项目。

　　与以往万达广场相比，番禺万博 CBD 商业广场体现了万达城市综合体升级的战略意图：

　　1. 空间布局的重新规划。与传统万达广场两庭一街孤立的空间相比，番禺万博 CBD 商业广场内部空间浑然一体，中庭与直街的连通，过桥的独特造型，使得内部空间更为独一无二。

　　2. 充分展示了万达广场主题的要求与格调的提升。当地是岭南文化的重要起源地，因此，淡雅的浅棕色，黑色曲线屋檐成为了主题体现的重要手段，淡灰淡棕的色调，勾勒出来的黑钢线，共同构成了番禺万博 CBD 商业广场的现代与地方、奢华与安静的美妙空间。

　　3. 人性化与安全的设计标准全面提升。作为番禺 CBD 商业区的高标准建造的项目，番禺万博 CBD 商业广场的生活设施配套大幅提升，除了结构与分区充分保障消防安全外，所有的主材如：抗菌纹理铝板、GRG 板材、金属面板等装饰材料，均能做到高档与安全兼备。是万达历来高标准项目的示范项目。

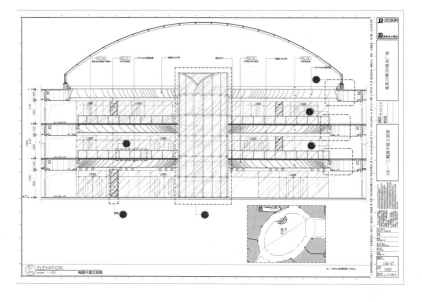

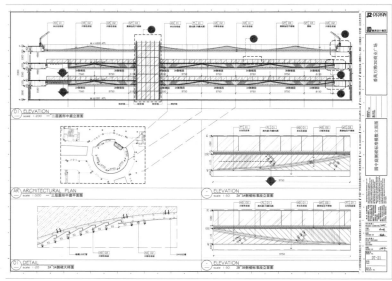

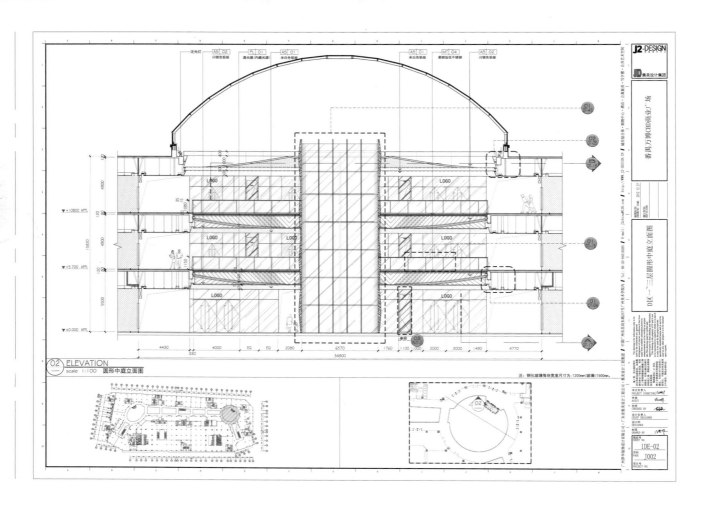

- New City Landmark — Mixed-use Architecture

Architects · J2-DESIGN

Dongguan, China

Dongguan Chang'an Wanda Plaza

东莞长安万达广场

Dongguan Chang'an Wanda Plaza is the key project of Dongguan, and is another masterpiece of Wanda Group after Guangzhou Baiyun Wanda Plaza for its layout in the Pearl River Delta.

In the design of large commercial interior decoration, designers are inspired by mixed "slash" in the project layout. As the main element of building exterior and the conception source of interior design, slash has become the link of the entire project. In cant modelling indoor areas are divided into five different modules and six different textures of materials, and through mixing, indoor has presented the visual effect of succinct modelling but rich change. This is in accord with the effect and atmosphere they pursuit on the design of this project.

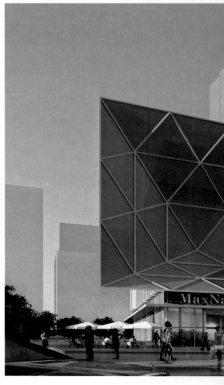

东莞长安万达广场项目由万达集团开发建设的东莞市重点项目,是继广州白云万达广场之后、万达集团布局珠三角的又一力作。

在巨大的商业中心内部装修的设计中,设计从该项目规划布局中混合的"斜线"获得了灵感。斜线作为建筑外部的主元素与室内设计的概念源泉,已成为连接整个项目的纽带。室内各区的斜面造型,分为5种不同的模型与6不同质感的材料,通过混搭,室内已呈现出简练造型但有着丰富变化的视觉效果。这正符合了设计师在这个项目上设计追求的效果、气氛。

• New City Landmark — Mixed-use Architecture

New City Landmark — Mixed-use Architecture

Design Agency: DCI Designgroup International
Client: Shanghai Electric Power Company Limited
Area: 305,890 m²

Shanghai, China

Yangshupu Power Plant Renovation Project

杨树浦发电厂综合改造项目

Yangshupu Power Plant renovation project located at the centre of Yanpu district of Shanghai. Along with the development of east bund, the first large sacle urban complex will be bulit in this area. Based on the development pattern of the entire Yanpu district cultural creative industry and insisted on the strategy of historic preservation as well as development, this project focus on create an overall development mode which combined shopping and traveling, and the development mode of industry fair and business travel. New energy themed industry fair fuse with investment promotion activities, business communication, trade, conference and forum.

The overall planning of this project, large sacle buildings are arranged along with Yangshupu road as the major urban interface with convenient traffic. Two 130m office buildings occupy the northwest of the site, act as the feature along with the L shape shopping center. The indoor shopping ceter and outdoor relaxing area form cyclic flow. Two five-star hotels with guest rooms of 250 and 300 occupy the east of the site, from north to south. The original steel structure and spatial scale of the old machine room are reserved. The room is transformed into function rooms of the hotel with full time restaurant, Chinese restaurant, feature restaurant, wine and cigar lounge as well as art gallery. The banquet and conference center close to the water side so as to extend for outdoor celebration activities. It is the theme function area of the hotel. The former dry coal shed on the southwest corner is transformed into new theatre and performance centerl which is flexible and able to meet various functional needs. Nearby giant steel structure building is transformed into vertical greenhouse. The distinguishing gaint steel structure of the building is reserved. A 60m high greenhouse with all kinds of plants is built within it. Walking upstairs there are tea room, cafe, DIY room, etc. Here you can have a rest, enjoying the varying scenery.

This project owns a perfect location with great views overlooking Huangpu River, However if the vast majority of the site elevation is exactly equals to sea

写字楼 OFFICE TOWER 80,640 SQ.M. HT: 129.9 M	酒店式公寓 SERVICE APARTMENT 33,790 SQ.M. L17-L36 HT: 152.7 M
商业 RETAIL 79,522 SQ.M. HT: 33 M	酒店A HOTEL A 26,502 SQ.M. L1-L16
变电站 SUBSTATION	酒店B HOTEL B 38,892 SQ.M. HT: 83.4 M
文化休闲 CULTURAL 5,581 SQ.M.	

level, plus the obstruction from 1.5m flood control wall, it will be negative for people's sight view. Therefore, by raising the main path and ground floor where people stay can ensure wide-field eyesight and make better use of the river scenery. The concept of three ground floor is proposed on that basis. The first floor is used to meet the needs of passenger traffic and freight transport. The second floor is used as major pedestrian passageway to each function area, providing space for outdoor activities at the same time. The roof layer and landscape constitute a huge sloping park.

杨树浦发电厂综合改造项目位于海派之都上海市杨浦区滨江核心带，坐拥城市副中心，随着东外滩的开发进程将打造此区域首座大型城市综合体，填补其无商圈的空白。基于整个杨浦区文化创意产业发展格局，坚持历史保护与功能开发并重的策略，在本案的规划策划方面则着重打造与购物旅游相结合的综合开发模式，以及工业博览与商务旅游的开发模式，以新能源为主题的工业博览会，并与招商活动，商务交流和交易，会议，论坛等融合。

在总体规划上，大体量的建筑被安排在了杨树浦路上用以塑造项目主要的城市界面，且在交通上具有优先通达性，两栋130m高的办公塔楼占据基地的西北角与L型沿界面展开的购物中心形成门户形象，其中室内的购物中心与室外休闲街区在人形动线上形成环通的路径；两座各拥有250间客房和300间客房的五星级酒店在基地的东侧由北向南排布，老气机房保留了原有钢结构和空间尺度，被改造成了酒店的功能配套，设有全日餐厅，中餐厅，特色餐厅，酒廊，雪茄吧，艺廊等，宴会及会议中心临近水岸的一端以便延伸户外的庆典活动，是酒店的主题功能区；西南角原来的干煤棚现做为新的剧场及演艺中心，可在空间的组成形式上灵活划分从而满足不同的功能需求；相邻的巨型钢架建筑被改造成了垂直温室，保留了其具有显著特征的外部巨型钢结构，在其内部构筑了一个60m高的温室花园，种植了各类植物上千种，由木栈道盘旋而上，沿途还穿插了茶室，咖啡，手工作坊等创意空间，可停下脚步在此休憩，体会沿途

变换的风景。

本案具有优越的黄浦江景观资源,然而若是绝大部分场地标高与海平面齐平,再加之又约 1.5m 高的防汛墙的阻挡,人的视野很难收纳到江景,将人们主要走动,停留和活动的地面层抬高则可保证视野的开阔性,大大增加了对江景资源的利用。于是便在此基础上大胆地提出了三个地面层的概念:首层作为车行解决客运及货运的需求;二层平台作为主要的人行动线可以无阻隔的到达各个功能区,同时提供了举办户外活动的空间;屋面层与景观地形串联共同构成巨大的坡地公园。

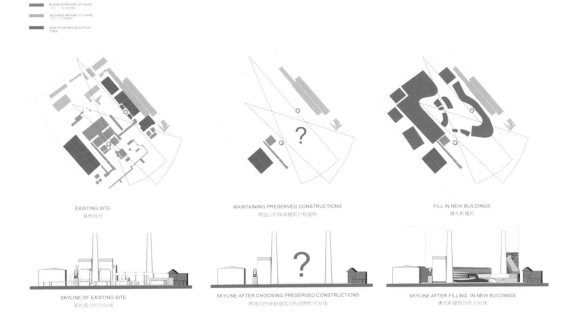

- New City Landmark — Mixed-use Architecture

New City Landmark — Mixed-use Architecture

Architect · Synarchitects
Program · Hotel, Office, Residential, Retail, Entertainment, Car Park

Ningbo, China

Ningbo Long Xing Plaza

宁波隆兴广场

The Long Xing Plaza was developed together with the clients in various steps.

In the progress the initial requirements were more and more consolidated and in an architectural language implemented. Located near the CBD of Ningbo at the former area of the Long Xing machine factory they did a conversion of the once industrial compound. It will consist of a 5 star hotel, offices, apartments, retail, very detailed pleasure grounds and a car park for 800 spaces.

The concept of the master plan is gathering the energy inside and simultaneously opens up towards the city. Facades angular to the streets articulate open spaces to welcome new visitors, curved buildings guiding the people into the center.

The north cluster is adapting to the neighboring buildings with perimeter block development (retail) to close the frontage. Seen from the city the area is maximized and at the client's request visually conceals an unattractive office-building.

The residential tower, 21 floors, is orientated with the lower part towards the residential district, with the upper part to the city. The three storey base of the tower consists of restaurants and retail. The southern located tower is merged out of two kinked slabs with 28 and 22 floors. Both are connected with a bridge that is five floors high. Merging the whole project the park gives the user a center to recreate. This part was exposed to the most changes. In the end a typical HOPSCA village came up, similar to the San Li Tun Village in Beijing.

建筑师在与客户共同商讨的基础上逐个开发隆兴广场的各个部分。

在这个过程中,最初的需求得以巩固和实现。在宁波中心商业区附近的隆兴机械厂,我们对这个曾经的工业场地进行了详谈。它将包括一个五星级酒店、写字楼、公寓、零售商场、非常详细的游乐场地及拥有800个泊车位的停车场。

总体规划的理念是集聚内部能量,同时向城市开放。朝向街道的尖角外立面连接欢迎新游客的开放空间,弧形的建筑引导人们进入中心。

北方的群集建筑适应周边建筑与周边的地块开发(零售),取消了临街面。从这座城市看,区域面积达到了最大,建筑师应客户的需求从视线上取消了一个没有吸引力的办公大楼。

21层的住宅大楼,较低部分朝向住宅区,较高部分朝向城市。大楼的底部三层包括餐厅和零售。位于南部的大楼中融入两栋分别为28楼和22楼的呈扭曲状的建筑。两者以一个5层高的天桥连接起来。公园融入整个项目,给用户一个用于重新创建的中心,这部分有着最大的变化。最后是一个典型的豪布斯卡村,类似于北京的三里屯。

New City Landmark — Mixed-use Architecture

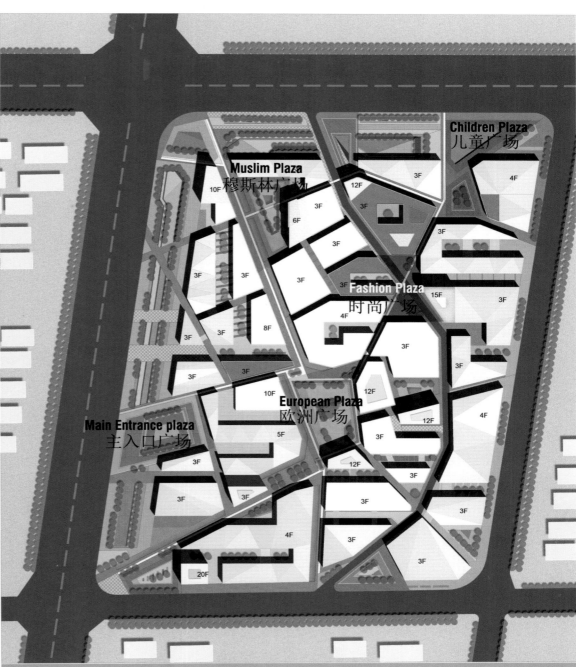

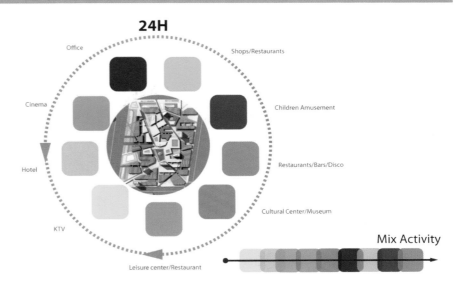

on the base of the original water channel, including reclaimed water, rainwater collection, etc, to make a little contribution to the city's water resource protection and play a positive role in energy conservation and emission reduction as well as ecology landscape of the project. Sustainable design, architectural physics and passive energy saving technology are applied in the architectural design of this project as well, in order to achieve all-round energy conservation and emission reduction.

阅海万家用地面积为 146 667m², 容积率 1.6 位于银川 CBD 的东侧, 老城区的北面。项目的目的是打造一个银川市的不夜城, 24 小时的文化商业休闲综合区。规划总平面套用一个旧城地图及叠加一个西夏国王拓拔元昊的印章, 并同时设计了文化广场、时尚广场、伊斯兰广场和商业娱乐广场。本项目在原有的水渠基础上设计水资源再利用系统, 包括中水利用, 雨水收集等, 为塞上湖城的水资源保护做一点贡献, 同时也为项目本身的节能减排及生态景观发挥作用。人文的可持续设计、建筑物理及被动式节能技术也同样应用到项目的建筑设计当中, 全方位实现节能减排。

• New City Landmark — Mixed-use Architecture

Architect · SURE Architecture
Client · Yinchuan C&D Group
Area · 144,000 m²

Yinchuan, Ningxia, China

Yinchuan Zhengyuanjie Development

银川正源街规划

Zhengyuanjie Development is a urban complex project with the area of 144,000 m², and 3.0 volume fraction. The project located at the east of Yichuan Yuehai CBD. In this project, the west of the site is commercial complex and the east of the site is residential area, while in between there is a business street. According to wind power and thermotechnical simulation, the commercial complex building is streamlined. The overhang scale of horizontal extension is designed according to the angle of sunlight in order to make better use of natural lighting as well as provide effective shading effect. The design of green roof reflects certain symbol of green architecture. Commericial complex is based on crosswise lines. In contrast, residential buildings emphasize on vertical lines to create personalized space.

　　正源街规划是一个城市综合体的项目，占地面积144 000m²，容积率3.0. 项目位于银川阅海CBD的东侧，正源街项目的正南方，一路之隔。本方案的西侧为商业综合体，东部是住宅区，中间为一条贯穿南北的商业街。通过风力及热工模拟，商业综合体部分设计成流线型，根据日照的角度设计横向外延的悬挑尺寸，使项目充分利用自然采光之余又可以有有效的遮阳作用。绿屋顶的设计也体现出一定的绿色建筑的特征。由于商业综合体以横向线条为主，因此，住宅采用相反的做法，强调竖向线条与商业成对比，从而打造个性化空间。

New City Landmark — Mixed-use Architecture

- New City Landmark — Mixed-use Architecture

Chengdu, China

Yangguang Chengdu

成都阳光新业中心

Architect · Laguarda.Low Architects, LLC
Client · Yangguang Co., Ltd.
Area · 88,000 m²

New City Landmark — Mixed-use Architecture

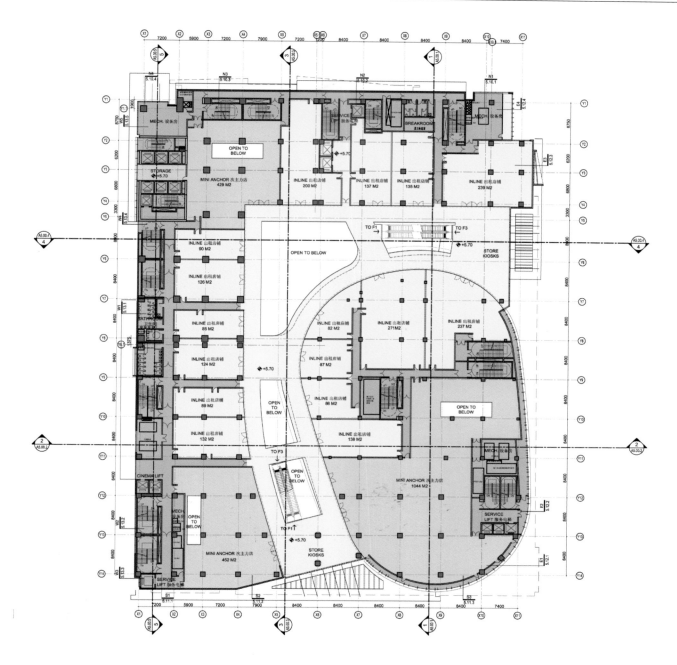

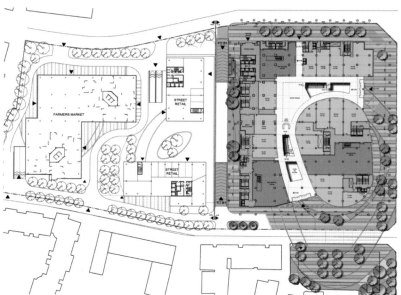

Located on the highly visible southeast street corner on the 3rd Ring Road in Chengdu, the Yangguang Chengdu posed the challenge of creating a dynamic environment that responded to the dense context and adjacent urban green while negotiating the project massing around pre-existing site conditions.

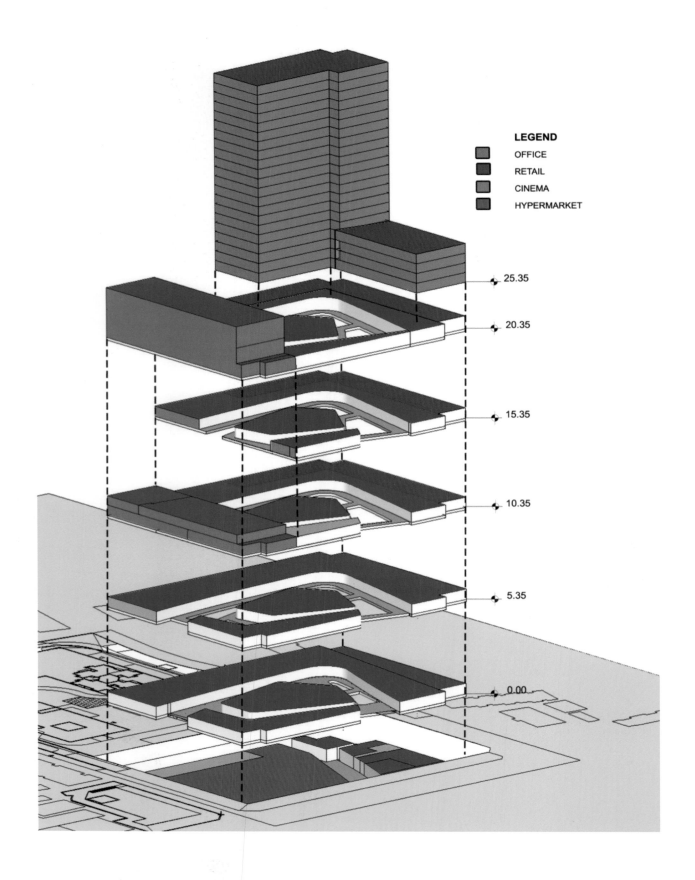

The key design gesture focused around the idea creating a structure evocative of a wooden puzzle box, that when the facades peel away the fenestration reveals an interior jewel-like curvilinear form that though faceting the light across its surface draws visitors deep within the space. The most unique aspect of the project generated by this system is manifested in a jewel-like object at the southeast corner of the project, representing the brand and identity for the entire development. The system, dynamic and flexible, utilizes the play between solid and void as a further response to linking the pre-existing massing with the new architecture.

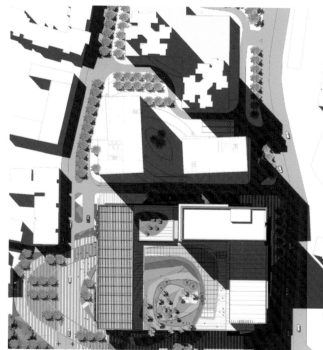

成都阳光新业中心位于成都三环的东南街角，位置极为醒目，这为创造一个充满活力的环境带来了挑战。本案不但要响应高密度的城市环境和附近的城市绿地，还要考虑到原有的基地条件，使项目体量不至于显得太突兀。关键的设计手法专注于创造一种能让人联想到木质魔术方块的结构体，若是剥开立面，室内如宝石一般圆润的造型便呈现于众人面前，光线照在曲面上再反射开去，璀璨动人，吸引游客愿意更深入地探索空间。本案最独特之处就是由这个系统所生成的，在项目东南角的宝石结构中分外耀眼，也成了整个开发项目的独特商标与标识。这个充满活力而又灵活的系统利用虚实空间的交错进一步呼应了原有体量与新建筑之间的衔接。

Contributors / 设计师名录

10 DESIGN

10 DESIGN is a leading international Partnership of Architectural, Urban, Landscape and Interior Designers and Computer Generated Imagery (CGI) specialists. They have spent the last two decades working in the emerging Asian and Middle Eastern markets and have established a reputation and brand name within the industry. The founding of 10 DESIGN was a result of the Partners' aspiration to create multicultural projects with architectural, economic, environmental and social integrity.

10 DESIGN works at a range of scales and operate within all sectors; including corporate, cultural, hospitality, retail, education and residential and have offices in Hong Kong, Shanghai, Edinburgh, and Dubai.

10 DESIGN has a group of specialist Landscape Designers who develop their concepts in conjunction with architects, arbourists, artists and the Sustainable Design Group. Collaboration and the sharing of ideas are the foundation of the group. A key focus of the Landscape Designers is the enhancement of communities and quality of life. They frequently work with the environmental specialists in projects that involve aspects such as Ecowalls or living green walls and other living elements.

The Sustainable Design Group research and utilise active and passive technologies and systems office wide that are specific to the Asian and Middle Eastern climates and environmental considerations such as anti-pollution measures. Their sustainable designers use the latest environmental simulation software including Ecotect and Vasari as well as Open Source software (OSS).

10 DESIGN has been in the World's Top 100 Architects since 2012.

5+design

Located in the heart of Hollywood, 5+design is a 90-person studio offering services in architecture, urban design, planning and interiors. They specialize in clients who desire innovation, and projects that create environments offering rich experiences to the user and the visitor. They are experts in retail, residential, mixed-use, hospitality and planning projects, and creating unique and compelling destinations for people worldwide. They are committed to their clients, striving always to provide design that is original, sustainable, and a pleasure to experience. Their work endeavors to find that which is unique and important within each project, creating an imaginative response to the context. They believe successful projects engage the user, opening up their sense of awareness, connecting them to a place to leave a lasting impression. The firm is led by five experienced partners who provide an inspirational foundation for a collaborative design studio of international designers from around the world. They are diverse, professional, and passionate. They speak over 12 languages and dialects, maintain offices in Hollywood and Shanghai and they are working with the world's most prestigious clients on projects that span four continents.

amphibianArc

amphibianArc is an architecture design firm headquartered in Los Angeles, with branch office in Shanghai, China. Founded by Nonchi Wang in 1992, the practice endeavors to synthesize between artistic expression and problem solving technique. Through a shared disclosure called Liquid Architecture, his work, encompasses not only the curvilinear forms prevalent in contemporary architectural scene, but also ideogramic methodology which is the foundation of Chinese word making.

Since its founding, amphibianArc has been designed a wide range of projects in both the U.S. and China, including the award winning Beijing Planetarium (AIA/LA awards, China Zian Tien Yo Award), Foshan Dongping New City Mass Transit Center (2011 MIPIM Architecture Review Future Project Awards), and Hongxing Macalline Furniture Flagship Store (2012 MIPIM Architecture Review Future Project Awards). The practice's work has also been featured in publications including Architectural Record, Dezeen, LA Architect, and World Architecture.

ARCODEC-COM Architectural Studio

In 2006, Arch. Ion Eremciuc, founded the architecture & design studio "ARCODEC-COM". "ARCODEC-COM" is an architectural studio specializing in residential, commercial and industrial architecture in domestic and international markets based in Chisinau, Republic of Moldova.

The Studio is providing architectural services using technologically superior processes, providing greater value for clients and enhanced design and construction, interior design, visualization services.

The target client is segmented into four categories; home owners, developers, government, and contractors. "ARCODEC-COM" Architecture's competitive edges are the knowledge of digital-based design resources.

ATENASTUDIO

ATENASTUDIO is a research and a design company based in Rome, working in the fields of architecture, urbanism, landscape and interior design. The studio was established by Marco Sardella and Rossana Atena in 2005.

ATENASTUDIO has been strongly involved on international level in several design projects and competitions with professionalism and thorough knowledge of many aspects conceiving architecture, advanced technologies, landscape, urbanism and interior design.

As part of a continuous process of research and academic involvement, ATENASTUDIO has been recognized on International level with many architectural lectures and publications. ATENASTUDIO has received many awards for winning competitions and prizes and had the opportunity to exhibit its work and research at various exhibitions in Italy, Spain, Germany and China.

BIG

BIG is a young architectural company, characterized by an entrepreneurial spirit, true team-work across expertise areas and new ways of approaching conventional tasks. They have an informal work environment where camaraderie and collegial support are highly valued and where ambition, very high work morale and dedication to being the innovators of our field unify the staff. Their firm is characterized by creativity, high energy and a unifying team spirit. Additionally, they are dedicated to creating and maintaining a cool workplace where they want to be and they continuously work at becoming better at what they do.

BIG currently comprises a group of architects, designers, and thinkers operating within the fields of architecture, urbanism, research, and development which are comprised of over 20 nationalities. The office is currently involved in a large number of projects throughout Europe, Asia and North America. BIG's architecture emerges out of a careful analysis of how contemporary life constantly evolves and changes, not least due to the influence of multicultural exchange, global economic flows and communication technologies that together require new ways of architectural and urban organization. In all their actions they try to move the focus from the little details to the BIG picture.

BLUA

Founded by Bin Lu, BLUA is a design group dedicated to explore the future of contemporary architecture. It is a crossover platform to transform arts (drawing, painting.etc), biology, and science into architecture. BLUA works across scales on design projects in a range of fields including architecture, urban design, industrial design, fashion design and advertising. Their ongoing design research involves the creation of generative design methodologies developed from a series way of using color.

Bin Lu, originally from Hangzhou, China studied under Zaha Hadid at the University of Applied Arts in Vienna, die Angewandte. He earned his master degree of architecture from Southern California Institute of Architecture (SCI-Arc) where he was an assistant teacher for the studio Tom Wiscombe. He also worked as a research assistant at the university of North Carolina (UNC), Digital Design Center. In 2007, he won the North Carolina Out-of-State Special Talent Architectural International Prize.

He was previously design director for EMERGENT in Los Angeles, specializes in digital design methodologies and scripting, right-hand to Principal Tom Wiscombe. He was in charge of various internationally renowned projects, including the Garak-Dong Wholesale Market, the Forrest Place Public Artwork, the Flower Street Bioreactor, the Land Art Generator, and the Shenyang National Sports Center. Most notably, Bin was Leader Designer for Shenyang Judo Arena, which has been hailed as one of the most important sports buildings of the 12th National Games of China. He also previously worked as a representative at KPF, MCM, RTKL and he is also the Director of China at DRDS.

B+H Architects

B+H is an award-winning, full-service architecture firm recognized for excellence in retail design, having worked with the top retail brands including Nike, Sephora and Ikea. Headquartered in Toronto, Canada, B+H was one of the first foreign firms to establish a presence in China, opening an office in Shanghai in 1992. Today in Asia, B+H has evolved into a full-service architecture practice with more than 200 employees, designing for retail, commercial, mixed-use, residential, healthcare, hospitality and transportation projects. In Asia, the firm has offices in Shanghai, Hong Kong, Beijing, Singapore and Ho Chi Minh City.

Buro Ole Scheeren

Buro Ole Scheeren is an international architecture firm that applies critical thinking to a process that extends beyond the boundaries of conventional architectural production.

German-born Ole Scheeren is principal of Buro Ole Scheeren and an internationally celebrated architect and visiting professor at Hong Kong University. Educated at the universities of Karlsruhe and Lausanne, he graduated from the Architectural Association in London and was awarded the RIBA Silver Medal.

At Buro-OS, Ole Scheeren is currently working on a series of projects, including Angkasa Raya, a 268 m tall landmark building in the center of Kuala Lumpur; a studio/gallery building for a Beijing-based artist; and DUO in Singapore. His investigation and exploration of new prototypes for architecture at various scales also includes the development of a new kind of kinetic performing art space/arena, as well as an 800,000m^2 mixed-use development in Chongqing. Furthermore, Ole Scheeren is working on the design for an arts center in Beijing and a contemporary art museum in central China.

COOP HIMMELB(L)AU

COOP HIMMELB(L)AU was founded in Vienna in 1968 and has since then been operating under the direction of CEO and design principal Wolf D. Prix in the fields of art, architecture, urban planning, and design. Another branch of the firm was opened in the United States in 1988 in Los Angeles. COOP HIMMELB(L)AU currently employs over 100 people from 19 different countries. In numerous countries the team has realized museums, concert halls, science and office buildings as well as residential buildings. Presently COOP HIMMELB(L)AU is working on various projects in Europe, Asia and the Middle East.

The company's most well-known international projects include the Falkestrasse attic conversion in Vienna, the multifunctional UFA Cinema Center in Dresden, the BMW Welt in Munich, the Akron Art Museum in Ohio, the Central Los Angeles Area High School #9 for the Visual and Performing Arts, the Busan Cinema Center in Korea and the Dalian International Conference Center in China. Among projects currently under construction are the Musée des Confluences in Lyon, France, the House of Music in Aalborg, Denmark, and the European Central Bank (ECB) in Frankfurt/Main, Germany.

Wolf D. Prix, born in 1942 in Vienna, is co-founder, Design Principal and CEO of COOP HIMMELB(L)AU. He studied architecture at the Vienna University of Technology, the Architectural Association of London as well as at the Southern California Institute of Architecture (SCI-Arc) in Los Angeles.

Wolf D. Prix is counted among the originators of the deconstructivist architecture movement. COOP HIMMELB(L)AU had its international breakthrough with the invitation to the exhibition „Deconstructivist Architecture" at MoMA New York in 1988. Over the years Wolf D. Prix/ COOP HIMMELB(L)AU was awarded with numerous international architecture awards.

Chapman Taylor

Chapman Taylor is an international practice of architects, masterplanners and designers established over 50 years ago in the UK. This international team of over 500 people operates from 17 regional offices, undertaking projects in over 80 countries worldwide.

Since 1959, they have established a reputation for delivering commercially successful, creative and innovative environments across a variety of sectors worldwide. Combining a strong ethos for high quality design with a deep understanding of commercial requirements enables them to deliver schemes that exceed their clients' expectations and provide award winning environments that people enjoy.

Chapman Taylor opened their Shanghai office in 2007, offering services across the following sectors: masterplanning; mixed use; marina & waterside planning and architectural design; retail; leisure; residential; workplace; hospitality; trans¬portation; interiors & graphics. Their primary goal is always to create exceptional design that is also highly commercially successful.

They design and deliver realistic and contextual urban masterplans which restore and improve the grain of our cities, or create new sustainable urban environments. As leaders in the field of urban regeneration they have created international award-winning schemes that respond to both the scale and needs of the built environment and the community.

As architects with unrivalled international experience of multi-use projects, their concepts are carefully crafted from an understanding of the changing needs, tastes and expectations of the city dweller. Their designs create a synergy between places and buildings, provide aesthetic inspiration, and make commercial and environmental sense.

It is their objective to continually evolve as an architectural practice, always striving to find new solutions to traditional challenges and to provide the best possible service to their clients and the community.

Decode Urbanism Office

Decode Urbanism Office(DUO), directed by Jianjunkai and Jianglong, is committed to designing and creating urban life experiences in different scales. They understood this life experience design not only limited in traditional design industry from products, furniture, devices and display design to the interior, building, site and landscape design. In Decode, they also pay close attention to developments in fields as diverse as sociology, psychology, information technology, manufacturing, materials science and new energy as they work to study their relation to urban life, improve the possibility of city life which is shown in the final form of a design product. DUO's attention crosses household utensils to urban space and takes the study object biof different scales as a system in common. Design is a process from finding the core of a problem to proposing a solution. No matter how complex the context is, DUO looks for the core behind the presentation and solve the problem with a method similar to decoding, for which is DECODE named.

Mr. Jian studied parametric urbanism in AA School, AADRL, and got his Master's degree in architecture there. He has been committed to studies on parametric design and its application in engineering. He worked at Zaha Hadid Architects Studio London for years. In 2009, Mr. Jian was shortlisted in the annual Evolo Competition and became the first Chinese award-winning designer at that competition.

Jiang Long attended the Architecture Department of TU Delft and received his Master's degree in architecture there. His graduation design received honors marks. Mr. Jiang had years of experience in working for Zaha Hadid Architects Studio London.

Design International

Founded in Toronto in 1965, Design International operated out of the American Market in its early days, but quickly developed into a major international player and was one of the first companies that truly worked on a global scale.

Design International's vision is to develop not only the character of the building, but also a sense of belonging for its customers to achieve a result that is fresh yet sympathetic to the local surroundings.

The company's portfolio includes award winning projects in over 45 countries and over 9,290,304m² of world class architecture."

Design International is an award winning architecture boutique, with a multi-disciplinary staff in 7 integrated divisions with a singular corporate commitment to create world class projects for their clients.

DCI Designgroup International

DCI Designgroup International (DCI) is an international architecture and design firm providing professional design service including urban complex design, shopping center design, department store design, hotel design, catering and entertainment design, office and residence design etc. Over the past two decades, DCI has accomplished over 200 projects in east Asia and China. DCI integrates international resources, senior specialists from New York headquarter, Asian central office in Shanghai and regional office in Beijing, as well as international consultants of professional field. This professional design team is now working on various design projects in different regions.

DCI has classic design team, concentrate on providing omnibearing professional services including overall planning, architectural design, commercial plan, interior design, brand design and environmental design. DCI is able to provide commercial plan and design assessment services on different stages, especially good at combining business strategy with urban complex design and reaching a win-win situation between development and operation, in order to meet the needs of Chinese Market

Ector Hoogstad Architecten

Ector Hoogstad Architects (EHA) is an independent all round architectural practice based in Rotterdam, headed by Joost Ector (design principal) and Max Pape (managing director). EHA has a leading position in the Netherlands' internationally highly ranked design-landscape. In over fifty years, the firm has built up an impressive portfolio of projects, remarkable both for its range and variety as well as its consistently high level of both architectural and technical quality. The firm has a large number of striking and significant buildings to its credit, such as the Netherlands Ministry of VROM (Spatial Planning and the Environment), several theatres, and buildings for various universities.

Ir. Joost Ector (1972), Partner and Design Principal, graduated cum laude in 1996 from Eindhoven University of Technology and then joined Hoogstad Architecten, where he became project architect in 1999 with his winning design for the HES Amsterdam. In 2001 he became a Director of the firm and in the following year became owner together with Max Pape of what is now Ector Hoogstad Architects (EHA).

F. Bozkurt Gürsoytrak / BOYUT ARCHITECTURE CO., LTD.

Bozkurt Gürsoytrak started his career working at several projects at Yalçın-Beate Oğuz Atelier, first as a trainee architect between 1976—1979 and then as an architect between 1979—1982. After completing Man Engine Factory project with Erhan Kocabıyıkolu and Faruk Eim between January to November in 1982, he did his military service. After doing his military service, again, he worked with Yalçın-Beate Oğuz between March 1983 and September 1983 in several projects. In 1983, he worked as a researcher in METU Faculty of Architecture.

He founded BOYUT ARCHITECTURE CO., LTD. on March in 1984. Besides working as a freelance architect since 1984, he also works as a part-time lecturer in Gazi University, Faculty of Engineering and Architecture since 2003.

His works sent to national and limited architectural competitions since he was a student at university were awarded with different degrees, including many first prizes. He also took place in different jury committees in several national architectural project competitions in various time periods.

He is a member of TMMOB Chamber of Architects, Turkish Freelance Architects Association, Association of Architects in 1972, The Union of Chambers and Commodity Exchanges of Turkey (TOBB) Technical Consultants Assembly and Urban Land Institute.

He mostly designs multipurpose complexes, trade centers, shopping malls, and tourism, office, housing, health and sport buildings.

FEMA Architectural Co. Ltd.

FEMA Architectural Co. Ltd. was founded in 1990. Since their establishment, they have been working with both private and public sector. In this process, many Multi-storey Buildings, Shopping Malls, Hotels, Educational Campuses, Cultural and Convention Centers, Sports Buildings and Houses have been designed. their firm provides services on project design and management, technical specifications, application advisory and consultancy, research and feasibility studies.

HENN

HENN is an internationally operating German architectural practice with more than 30 years of building expertise in the fields of culture and administration buildings, education, research and development and production buildings.

Projects such as Die Autostadt, Die Gläserne Manufaktur and the Project House in the Research and Innovation Center of BMW have been internationally acclaimed; ongoing large-scale projects in China include the headquarters for the two biggest life insurance companies China Life and Taikang Life, a production plant for BMW in Shenyang and the Science City of Nanopolis in Suzhou.

As a general planning practice HENN provides experience in all work phases. The broad scope of work includes architectural planning, interior design, masterplanning, quantity surveying, construction management and LEED certification.

The office is managed by Gunter Henn and nine partners. These days, 330 architects, designers, planners and engineers work in project teams in their offices in Munich, Berlin, Beijing and Shanghai.

HMA Architects & Designers

HMA is an international design company taking the design of commercial building and creative cultural industry park as the core business.

In recent years, HMA has provided the representative design experience for peers in the comprehensive protect of the industrial heritage and the design of creative cultural industry park. And it has completed "Shanghai No. 8 Bridge", "EXPO2010 Urban Best Practice Area B4 Pavilion", "Shanghai Daning Center Plaza" and so on. With its professional and creativity, HMA has been the pioneer and leader recognized by the professions in the domestic building industry.

In the field of commercial real estate, the efforts of the HMA also won a lot of attention and recognition. With the persistent pursuit of "providing consumers with new consumption experience", HMA successful completed the design of urban commercial complex projects for a series of water swimming cities such as Nanjing, Tianjin and Panjin, and made a significant contribution in creating characteristic commercial pedestrian street.

HMA has contributed their unique creation to the development of the Chinese commercial real estate and creative culture industrial park, and committed to create more social value and rich experience of multidimensional space for city, developers and urban residents by creativity and design.

Holm Architecture Office (HAO)

HAO was founded in 2010 and is based in Brooklyn, NYC. The HAO office focuses on architectural solutions based on cross-disciplinary and programmatic thinking. The office works with a wide range of collaborators, seeking new design approaches for the areas of architecture, art and design.

HAO has worked locally and globally on projects ranging in scale from urban art installations to master plans. HAO recently completed the first phase of the re-design of the Coleman Oval Skate Park and Park, situated under the Manhattan Bridge, in collaboration with Architecture for Humanity and Nike.

HAO is currently involved in the development of three master plans in China and In April of 2013, the Olympic Samaranch Memorial Museum opened in Tianjin, China, a 20,000 m² project designed by HAO that was nominated for the prestigious WAF / World Architecture Festival.

HWCD

HWCD (Harmony World Consultant & Design) is an International Consultant and Design Company with offices in Shanghai, London and Barcelona. Its dedicated team of architects, landscape and interior designers offers a professional, personalized and comprehensive design service.

Focused on creating environments that affect people's lives, the company has developed cultural-themed projects, boutique hotels and residential and mixed-use projects not only in China but also in Europe.

HWCD's partners come from different countries and have different backgrounds. They work together in a constantly changing environment taking advantage of the large Chinese tradition but always opened to the new trends.

J2-DESIGN

J2-DESIGN is a successful design brand in Asia. As the core enterprise of Jimei Group, J2 brings professionals of Guangzhou and Hong Kong together and has been the most representative trans-regional design team in the South of China. Today, J2's interior design of "urban complex" with hotel, business center, apartment and office building as the main body has become the representative of industry. Its customers include many well-known enterprises: Vanke, Wanda, Longfor, Accor, Wyndham and so on.

J2-DESIGN takes upgrading customers' industrial value as its priority. It not only has rich imagination and steady management experience in the field of interior design; J2 can also provide the leading professional services in the planning of projects which cross the boundary, architectural landscape (J2 collaborates with Denmark's biggest architectural design agency, VLA.), system integration of visual arts themed decoration.

J2-DESIGN is a successful Asian design brand. As the core enterprise of Jimei Group, and has a number of design elites which embrace an international perspective; with its strong comprehensive design ability and professional project management system, J2, one of the biggest design team in China, focuses on the field of "urban complex" with hotel, business center, apartment and office building.

J2-DESIGN regards enhancing customers' brand value as its mission. It has gradually received customers' recognition with its imaginative creativity and steady design quality for many years; J2 also shares creative resources and project management experience with international teams in the fields of architectural landscape design (J2 collaborates with Denmark's biggest architectural design agency, VLA.) , lighting design, visual communication art, and soft adornment design through their strategic cooperation, which brings J2's surprising ability of project integration and planning.

JDS Architects

Julien De Smedt is the founder and director of JDS Architects based in Brussels, Copenhagen, Belo Horizonte and Shanghai. Julien's commitment to the exploration of contemporary architecture has helped to re-energize the discussion of the practice with projects such as the VM Housing Complex, the Mountain Dwellings, the Maritime Youth House and the Holmenkollen Ski Jump.

Prior to founding JDS Architects, Julien worked with OMA/Rem Koolhaas, Rotterdam and co-founded and directed with Bjarke Ingels the architecture firm PLOT in Copenhagen.

Among other awards and recognitions, Julien received the Henning Larsen Prize in 2003 and an Eckersberg medal in 2005. In 2004 the Stavanger Concert Hall was appointed world's Best Concert Hall at the Venice Biennale, and the Maritime Youth House won the AR+D award in London and was nominated for the Mies van der Rohe award. In 2009, Julien De Smedt received the Maaskant prize of Architecture, and in 2011 he received the WAN- World Architecture News '21 for 21' Award – leading architects of the 21st century.

KaziaLi Design Collaborative

After years of practicing architecture throughout the United States and around the globe with international firms HOK and SmithGroup, Clay Vogel formed Kazia Design Collaborative in 2006. Based in Chicago, Kazia Design Collaborative found quick and lasting success in China's growing market. After two years of continual collaboration with Li Chunguang in China, Clay Vogel and Li Chunguang founded KaziaLi Design Collaborative (KaziaLi).

They are a unique design firm that has true integration between our United States and China bases. Their research-based, cross-cultural design approach has allowed, for the past five years, to design and complete successful large-scale projects within China.

As a young international design company, their interests lie in creating great buildings which exceed their clients' expectations. In so doing they are specifically interested in defining a memorable and appropriate form for the building: a form which solves the program – satisfying the questions of spatial organization and functional adjacencies, budget, schedule, and technology – in an elegant and refined manner, yet within a framework that is responsive to the local culture, the urban or suburban patterns, the natural environment, and the socioeconomic setting.

They believe a building's details are just as important as a powerful, overall vision, and they are innovative and cutting-edge in the use of technologies in their designs, as well as in the organization and delivery of services. Their projects include hospitals, museums, hotels, office buildings, educational facilities, luxury villas and clubs, and range in scale from thousands to millions of square meters. They actively nurture cooperative relationships with local firms in the global environment, and use our knowledge of design and technical expertise to provide insight and support to the native company. They create unique solutions for challenging commissions for global clients in the US and China, as well as the Middle East, India, Korea, and Vietnam.

Laguarda.Low Architects, LLC

Based in Dallas, Texas, USA, Laguarda.Low Architects, LLC is an award-winning international architectural practice founded in the year 2000 by a closely knit group of design-oriented architects with many years of shared knowledge and experience. The practice has worked extensively abroad in Europe, Asia and North and South America on developments of all types, ranging from master plans, large and small-scale mixed use developments, retail/entertainment centers, office buildings, resorts, residential developments and public projects. Always evolving to meet the needs of its international clients, the firm has opened a satellite office in Beijing, China and an affiliate office in Tokyo, Laguarda.Low+Tanamachi.

Design is the focus of Laguarda.Low's practice. The success of the firm in the international marketplace is a testament to their ability to respond efficiently and effectively to demanding client needs and to deliver innovative design solutions that add value to commercial development in a timely fashion. Laguarda.Low's design philosophy is open and continually developing. Their methodology focuses each architectural project on the understanding and development of its urban design role, and their numerous repeat clients are a testament to the success of this approach. Laguarda.Low's designs add value.

Laguarda.Low adheres to the policy that good clients create good projects, and is therefore intent on retaining clients by combining creative value-added design with excellent service. Among the global clients for which Laguarda.Low has provided its innovative design services include: Europe: Sonae Imobiliaria, Group Lar Grosvenor, Eroski, ING, Global Trade Centre (GTC), Reform Sp.Zo.o., PROCOM, South America: Sonae Enplanta, Sonae Sierra, Grupo Multiplan, Asia: Mitsui Fudosan, Daiei, Daiwa House, Diamond City, Cosmo Oil Company, Mitsubishi Chemical, Tokyu Group, Tokyo Tatemono Cosco Real Estate, Zenith, Core Pacific Group, China Resources Land, Shenzhen- New World Group, Overseas Chinese Town (OCT). North America: City of Dallas, Westfield, General Growth Properties.

Laguarda.Low utilizes the latest in global communications technology to work successfully around the globe. Some of the tools currently utilized include secure FTP sites for file transfers, a dedicated in-house video conferencing system, high-speed internet connections, global cellular telephones and globe-trotting computer laptops. To be successful in the world marketplace, the ability to work virtually in any location at any time seamlessly and without interruption with all clients is recognized and adds to the flexible, nimble nature of Laguarda.Low's practice.

In the last 10 years, Laguarda.Low has executed more than 300 projects in over 25 countries, ranging from the United States, China, South Korea, Japan, India, Indonesia, Australia, Brazil, Chile, Mexico, Spain, Italy, Portugal, Germany, Greece, Russia, Ukraine, Poland, Bulgaria, Hungary, Croatia, Serbia, Macedonia, Albania, to Kazakhstan and Romania. Project types that Laguarda.Low includes in its area of expertise are both small and large-scale Mixed-use, Retail/Entertainment, Hotel, Office, Institutional and Residential developments. Built projects by Laguarda.Low range from one of the largest retail centers in South America to the prototypical Cosmo gas station in Yokohama, Japan. The breadth and range of their work is constantly growing.

LAVA

Chris Bosse, Tobias Wallisser and Alexander Rieck founded multinational form, Laboratory for Visionary Architecture [LAVA], in 2007 as a network of creative minds with a research and design focus and with offices in Sydney, Shanghai, Stuttgart and Abu Dhabi.

LAVA explores frontiers that merge future technologies with the patterns of organization found in nature and believes this will result in a smarter friendlier, more socially and environmentally responsible future.

The potential for naturally evolving systems such as snowflakes, spider webs and soap bubbles for new building typologies and structures has continued to fascinate LAVA – the geometries in nature create both efficiency and beauty. But above all the human is the centre of their investigations.

Structure, material and building skin are three areas LAVA believes that architecture can learn so much from nature. Projects incorporate intelligent systems and skins that can react to external influences such as air pressure temperature, humidity, solar-radiation and pollution.

LAVA has designed everything from pop up installations to master plans and unban centers, from homes made out of PET bottles to retrofitting aging 60s icons, from furniture to hotels, houses and airports of the nature.

LAVA won an international competition to design the centre of Masdar, the world's first zero carbon city in the UAE.

LWK & Partners

LWK & Partners is a versatile, multi-skilled practice that values originality of thought. They believe that good design is an intelligent response to every challenge. The solutions provided are unique and appropriate, unconstrained by a distinct aesthetic style. LWKP's works demonstrate their fundamental principles that design is a well though-out solution and is practical.

The design of the Guiyang Twin Towers has been a collaborative effort between LWK & Partners' Hong Kong and Shenzhen offices. This project's international team of designers represents many different countries, cultures and a diverse set of design backgrounds. Each team member brings something unique to the table, which makes for a design process which is always evolving as a product of investigation and group discussion.

Llewelyn Davies

Llewelyn Davies has been established in Hong Kong for over 25 years, operating as an international multi-disciplinary design practice encompassing Architecture, Planning, Landscape and Interior Design. The firm is also the holding company of Percy Thomas Partnership (HK) Ltd (PTP) which is part of the Percy Thomas Group founded in the United Kingdom more than 100 years ago.

Through the ownership of PTP, Llewelyn Davies has forged a strategic association with Capita Percy Thomas, the firm responsible for London's 2010 Olympic Games winning bid, which is also the fifth largest UK architectural Practice. Capita Percy Thomas is part of the Capita Symonds Group, a public listed Company in the UK with a staff strength of about 4,000 having a range of professional consultancy services worldwide.

To strengthen the alliance of Llewelyn Davies Ken Yeang Ltd. (United Kingdom) and the sister companies of Llewelyn Davies HK Ltd. (Hong Kong, china), Hamzah & Yeang (Malaysia), and North China Hamzah Yeang Architectural and Engineering Company (Beijing, Shanghai, Shenzhen, Guangzhou), and to reinforce the shared spirit and identity forged on the legacy of Llewelyn Davies founded more than 50 years ago in UK, the Hong Kong office of Llewelyn Davies, formerly trading under the name of LD Asia, shall now operate under the name of Llewelyn Davies.

The alliance, comprising over 400 staffs, continues to serve their clients globally, delivering projects spanning commercial, residential, healthcare, rail, aviation, education and projects in other specialized sectors, including ecomasterplanning and ecosystems consultancies.

MANUELLE GAUTRAND

Manuelle Gautrand was born on July 14, 1961 in Marseille (France). She obtained her graduate diploma in Architecture from the "Ecole Nationale Supérieure d'Architecture de Montpellier" in 1985. She worked for 6 year in different architecture studios in Paris. She founded her office in 1991, first in Lyons and then in Paris. She lives and works in Paris since 1994.

She is the principal architect and director of the agency MANUELLE GAUTRAND ARCHITECTURE. She mainly designs buildings in areas as diverse as cultural facilities (theaters, museums, and cultural centers), office buildings, housing, commercial and leisure facilities, etc.

Her clients are public contracting authorities as well as private firms, in France and abroad. In 2007 Manuelle Gautrand's "C42" Citroen Flagship Showroom on the Champs-Elysées Avenue in Paris gained attention and widespread acclaim in the international arena and from a large audience.

Perkins Eastman

Perkins Eastman is among the top design and architecture firms in the world. With 700 employees in 13 locations around the globe, Perkins Eastman practices at every scale of the built environment. From niche buildings to complex projects that enrich whole communities, the firm's portfolio reflects a dedication to progressive and inventive design that enhances the quality of the human experience. The firm's portfolio includes high-end residential, commercial, hotels, retail, office buildings, and corporate interiors, to schools, hospitals, museums, senior living, and public sector facilities. In 2011, Perkins Eastman merged with Ehrenkrantz Eckstut & Kuhn Architects (EE&K), significantly strengthening both practices. Perkins Eastman provides award-winning design through its offices in North America (New York, NY; Boston, MA; Charlotte, NC; Chicago, IL; Pittsburgh, PA; San Francisco, CA; Stamford, CT; Toronto, Canada; and Washington, DC); South America (Guayaquil, Ecuador); North Africa and Middle East (Dubai, UAE); and Asia (Mumbai, India, and Shanghai, China).

Progetto CMR

Progetto CMR, founded in 1994 by Massimo Roj, is an architectural consultancy firm skilled in Architectural and Building Services Engineering integrated design, set up by professional consultants who gained a worldwide experience in these fields.

Progetto CMR headquarters are based in Milan, while other offices are located in Rome, Beijing, Shanghai, Tianjin, Athens, Barcelona, Chennai, Dubai, Istanbul, Mexico City, Prague. Progetto CMR is the Italian member of European Architects Network.

Progetto CMR, UNI EN ISO 9001:2008 certified, is structured in six departments: Architecture, Engineering, Safety Management, Process Management, Industrial Design.

Progetto CMR Beijing started operating in the Chinese market in 2003 and now has three operative offices in Beijing, Shanghai and Tianjin, relying upon a network of domestic and international partners distributed all over the Chinese territory.

Progetto CMR (Beijing)'s design team is made up of both Italian and Chinese architects, cooperating at all levels of the design process to provide innovative design solutions tailored to the needs of the Chinese market. Thanks to a multicultural and multi-disciplinary approach, our designers are able to provide consultancies for urban planning, masterplan, landscaping, building design, interior design, and space planning solutions for residential, hotel, office, industrial, retails and public spaces.

RTKL Associates Inc.

RTKL is a worldwide architecture, engineering, planning and creative services organization. Part of the ARCADIS global network since 2007, RTKL specializes in providing its multi-disciplinary services across the full development cycle to create places of distinction and designs of lasting value. RTKL works with commercial, workplace, public and healthcare clients on projects around the globe.

SDA | Synthesis Design + Architecture

Synthesis is an emerging contemporary design practice with over 20 years of collective professional experience in the fields of architecture, infrastructure, interiors, installations, exhibitions, furniture, and product design. Founded in 2011, Synthesis work has already begun to achieve international recognition for its design excellence.

The diverse team of multidisciplinary design professionals includes registered architects, architectural designers and computational specialists educated, trained, and raised in the USA, UK, Denmark, Portugal, China, Taiwan, China, Canada, Iran and Jordan. This diverse cultural and disciplinary background has supported their expanding portfolio of international projects in the USA, Canada, UK, Russia, Thailand, Korea, China.

Synthesis' work focuses on the merging of creative, intellectual, and technological design processes with intelligent fabrication and construction techniques to create beautifully crafted designs. They see each project as a unique opportunity to integrate diverse and often contending conceptions and constraints into a coherent whole.

SPARK

SPARK is an award-winning international architectural and design consultancy with proven expertise in architecture, urban design, landscape architecture and interior design. Foundeds in 2008, SPARK creates distinctive projects across Asia, Europe and the Middle East. Boasting a dynamic team of over 100 employees spanning 16 nationalities and three continents, SPARK combines the best experience of international and local talent. Driven by a philosophical approach to create architecture that is pragmatic, social and convivial, SPARK works closely with clients to create sustainable architecture that is underpinned by financial viability and the desire to improve the quality of life for all. With a presence in Beijing, London, Shanghai, Kuala Lumpur, Singapore and Abu Dhabi, SPARK's award winning projects include the rejuvenated Clarke Quay in Singapore, the Shanghai International Cruise Terminal, Starhill Gallery Kuala Lumpur and the Raffles City projects in Ningbo and Beijing.

Synarchitects / Yingxi Zou

Yingxi Zou was born in 1973 in Zhengzhou, China. From 1993-1997, he Studied on Environmental art design, Central Institute of Arts and Crafts (Now: Academy of Fine Arts of Tsinghua University). He won the gold prize of second term Indoor Architectural Design Exhibition in 1999. In 2000, he went to Germany and enrolled at the University of fine arts Berlin. In 2004, he got Diploma, Architectural Studies, University of fine arts Berlin. Same year, he founded Synarchitects in Berlin. He joined World Chinese Union of Architects as a founder. Since 2004, he work as a master architect for Beijing Longanhuacheng Architectural Designing Co.,Ltd., China.

SURE Architecture

SURE Architecture was founded in London, England, and is an international design company which focuses on the research and practice of sustainable urban renewal and ecological architecture. The existing London Company, as the general headquarter, has set up regional offices respectively in Beijing and Hong Kong, China, hosting business for Asia and the Pacific region. SURE Architecture owns a strong international team from different backgrounds on design and research. Sustainable urban renewal and ecological architecture is their core theory and design method. Service includes project planning orientation, urban planning, architectural design, landscape design, interior design, curtain wall design, ecological consultant and project management, etc.

The Jerde Partnership, Inc

The Jerde Partnership is a visionary architecture and urban design firm that creates dynamic places that deliver memorable experiences and attract over 1 billion people annually. Founded in 1977, the firm has pioneered "placemaking" throughout the world with projects that provide lasting social, cultural and economic value and promote further investment and revitalization. Based in a design studio in Los Angeles with project offices in Shanghai, Hong Kong, Seoul, and Berlin, Jerde takes a signature, co-creative approach to design and collaborates with private developers, city officials, specialty designers and local executive architects to realize the vision of each project. The firm has received critical acclaim from the American Institute of Architects, Progressive Architecture, American Planning Association, International Council of Shopping Centers, and Urban Land Institute. To date, over 110 Jerde Places have opened in diverse cities, including Atlanta, Budapest, Hong Kong, Istanbul, Las Vegas, Los Angeles, Osaka, Rotterdam, Seoul, Shanghai, Tokyo and Warsaw.

UNStudio

UNStudio, founded in 1988 by Ben van Berkel and Caroline Bos, is a Dutch architectural design studio specializing in architecture, urban development and infrastructural projects. The name, UNStudio, stands for United Network Studio, referring to the collaborative nature of the practice. In 2009 UNStudio Asia was established, with its first office located in Shanghai, China. UNStudio Asia is a full daughter of UNStudio and is intricately connected to UNStudio Amsterdam. Initially serving to facilitate the design process for the Raffles City project in Hangzhou, UNStudio Asia has expanded into a full-service design office with a multinational team of all-round and specialist architects.

ARTPOWER

Acknowledgements

We would like to thank all the designers and companies who made significant contributions to the compilation of this book. Without them, this project would not have been possible. We would also like to thank many others whose names did not appear on the credits, but made specific input and support for the project from beginning to end.

Future Editions

If you would like to contribute to the next edition of Artpower, please email us your details to: artpower@artpower.com.cn